i-Smooth Analysis

Scrivener Publishing
100 Cummings Center, Suite 541J
Beverly, MA 01915-6106

Publishers at Scrivener
Martin Scrivener(martin@scrivenerpublishing.com)
Phillip Carmical (pcarmical@scrivenerpublishing.com)

i-Smooth Analysis

Theory and Applications

A.V. Kim

Scrivener Publishing

WILEY

Co-published by John Wiley & Sons, Inc. Hoboken, New Jersey, and Scrivener Publishing LLC, Salem, Massachusetts.
Published simultaneously in Canada.

Cover design by Kris Hackerott

Library of Congress Cataloging-in-Publication Data:

ISBN 978-1-118-99836-6

Printed in the United States of America

10 9 8 7 6 5 4 3 2 1

To sweet memory of my brother Vassilii

Contents

Preface

i-Smooth analysis is the branch of functional analysis, that considers the theory and applications of the invariant derivatives of functions and functionals.

The present book includes two parts of the *i*-smooth analysis theory. The first part presents the theory of the invariant derivatives of functionals.

The second part of the *i*-smooth analysis is the theory of the invariant derivatives of functions.

Until now, *i*-smooth analysis has been developed mainly to apply to the theory of functional differential equations. The corresponding results are summarized in the books [17], [18], [19] and [22].

This edition is an attempt to present *i*-smooth analysis as a branch of the functional analysis.

There are two classic notions of generalized derivatives in mathematics:

a) The Sobolev generalized derivative of functions [34], [35];
b) The generalized derivative of the distribution theory [32].

In works [17], [18], [19] and [25] the notion of the invariant derivative (*i*-derivative) of nonlinear functionals was introduced, developed the corresponding *i*-smooth calculus of functionals and showed that for linear continuous functionals the invariant derivative coincides with the generalized derivative of the distribution theory.

The theory is based on the notion and constructions of the invariant derivatives of functionals that was introduced around 1980.

Beginning with the first relevant publication in this direction there arose two questions:

Question A1: *Is it possible to introduce a notion of the invariant derivative of functions?*

Question A2: *Is the invariant derivative of functions concerned with the Sobolev generalized derivative?*

This book arose as a result of searching for answers to the *questions **A1** & **A2***: there were found positive answers on both these questions and the corresponding theory is presented in the second part.

Another question that initiated the idea for this EDITION was the following

Question B: *Does anything besides a terminological and mathematical relation between the Sobolev generalized derivative of functions and the generalized derivative of distributions?*

At first glance the question looks incorrect, because the first derivative concerns the finite dimensional functions whereas the second one applies to functional objects (distributions – linear continuous functionals).

Nevertheless as it is shown in the second part the answer to the *question **B*** came out positive: the mathematical relation between both generalized derivatives can be established by means of the invariant derivative.

One of the main goals writing this book was to clarify the nature of the invariant derivatives and their status in the present system of known derivatives. By this reason we do not pay much attention to applications of the invariant derivatives and concentrate on developing the theory.

The edition is not a textbook and is assigned for specialists, so statements and constructions regarding standard mathematical courses are used without justification or additional comments. Proofs of some new propositions contain only key moments if rest of the details are obvious.

Though the edition is not a textbook, the material is appropriate for graduate students of mathematical departments and be interesting for engineers and physicists. Throughout the book generally accepted notation of the functional analysis is used and new notation is used only for the latest notions.

Acknowledgements. At the initial stage of developing the invariant derivative theory the support of the professor V. K. Ivanov had been very important for me: during personal discussions and at his department seminars of various aspects of the theory were discussed. Because of his recommendations and submissions my first works on the matter were published.

At the end of the 1970-s and the beginning of the 1980-s, many questions were cleared up in discussions with my friends and colleagues: PhD-students Alexander Babenko, Alexander Zaslavskii and Alexander Ustyuzhanin. Theory of numerical methods for solving functional differential equations (FDE) based on *i*-smooth analysis was developed in cooperation with Dr. V. Pimenov[1].

The attention and support of professor A. D. Myshkis was important to me during a critical stage of *i*-smooth analysis development.

I am thankful to Dr. U. A. Alekseeva, professor V. V. Arestov, professor A. G. Babenko, professor Neville J. Ford for their familiarization with the preliminary versions of the books and useful comments and recommendations.

The author is thankful to a book editor for the great work on book improvement and to A. V. Ivanov for preparation of the printing version of the book.

Research was supported by the Russian Foundation for Basic Research (projects 08-01-00141, 14-01-00065, 14-01- 00477, 13-01-00110), the program "Fundamental Sciences for Medicine" of the Presidium of the Russian Academy of Sciences, the Ural-Siberia interdisciplinary project.

1 The author developed a general approach to elaborating numerical methods for FDE and Dr. V. Pimenov developed complete theory, presented in the second chapter of this book

Part I

INVARIANT DERIVATIVES OF FUNCTIONALS AND NUMERICAL METHODS FOR FUNCTIONAL DIFFERENTIAL EQUATIONS

Chapter 1

The invariant derivative of functionals

In this chapter we consider the basic constructions of nonlinear i-smooth calculus of nonlinear on $C[a,b]$ functionals.

A study of nonlinear mappings can be realized by local approximations of nonlinear operators by linear operators. Corresponding linear approximations are called derivatives of nonlinear mappings. Depending on the form of linear approximations of various types of derivatives can be introduced. Further

$C[a,b]$ is the space of continuous functions $\phi(\cdot):[a,b]\to \mathbf{R}$ with the norm $\|\phi(\cdot)\|_C = \max\limits_{a\le x\le b}\|\phi(x)\|$;

$C^k[a,b]$ is the space of k–times continuous differentiable functions $\phi(\cdot):[a,b]\to\mathbf{R}$;

$C^\infty[a,b]$ is the space of infinitely differentiable functions $\phi(\cdot):[a,b]\to\mathbf{R}$.

1 Functional derivatives

In the general case *a derivative* of a mapping

$$f: X \to Y \tag{1.1}$$

(X & Y are topological vector spaces) at a point $x_0 \in X$ is a linear mapping

$$f'(x_0) : X \to Y \tag{1.2}$$

which approximates in an appropriate sense the difference

$$f(x_0 + h) - f(x_0) \tag{1.3}$$

by h. Subject to the specific form of difference approximation difference(1.3) one can obtain various notions of derivatives.

Consider the classic derivatives of functionals [1]

$$V[\cdot] : C[a,b] \to \mathbf{R} \,. \tag{1.4}$$

1.1 The Frechet derivative

The Frechet derivative (strong derivative) of the functional (1.4) at a point $\phi(\cdot) \in C[a,b]$ is a *linear* continuous functional

$$L_\phi[\cdot] : C[a,b] \to \mathbf{R} \,, \tag{1.5}$$

satisfying the condition

$$V[\phi + \psi] = V[\phi] + L_\phi[\psi] + o(\psi) \,, \tag{1.6}$$

where $\displaystyle\lim_{\|\psi\|_C \to 0} \frac{\|o(\psi)\|_C}{\|\psi\|_C} = 0 \,.$

If there exists a functional L, satisfying the above conditions, then it is denoted by $\psi \to V'[\phi]\,\psi$ and is called the *Frechet differential.*

1.2 The Gateaux derivative

The *Gateaux derivative* of the functional (1.4) at a point $\phi \in C[a,b]$ is a linear mapping $V'_\Gamma[\phi] : C[a,b] \to \mathbf{R}$, satisfying the condition

$$V[\phi + \psi] - V[\phi] = V'_\Gamma[\phi]\,\psi + \varepsilon(\psi) \,, \tag{1.7}$$

where $\displaystyle\lim_{t \to 0} \frac{\varepsilon(t\,\psi)}{t} = 0 \,.$

[1] Complete theory see for example in [16], [28].

2 Classification of functionals on $C[a,b]$

Many specific classes of functionals have integral forms. The investigation of such integral functionals formed the basis of the general functional analysis theory.

Along with integral functionals (which are called *regular functionals*) beginning with the works of Dirac and Schwartz mathematicians we generally use *singular functionals* among which the first and most well known is the δ–function.

Further as a rule integrals are understandood in the Riemann sense [2].

2.1 Regular functionals

Analysing the structure of specific functionals one can single out basic (elementary) types of *integral* functionals on $C[a,b]$:

$$V[\phi] = \int_a^b \alpha[\phi(x)]\, dx\,, \tag{2.1}$$

$$V[\phi] = \int_a^b \beta[x, \phi(x)]\, dx\,, \tag{2.2}$$

$$V[\phi] = \int_a^b \beta[x, \int_x^b \phi(\xi)\, d\xi\,]\, dx\,, \tag{2.3}$$

$$V[\phi] = \int_a^b \int_a^b \omega[\phi(x), \phi(\xi)]\, dx\, d\xi\,, \tag{2.4}$$

$\alpha : \mathbf{R} \to \mathbf{R}$, $\beta : [a,b] \times \mathbf{R} \to \mathbf{R}$, $\omega : \mathbf{R} \times \mathbf{R} \to \mathbf{R}$ are continuous functions.

[2] In some examples we use the Stilties integral.

Special classes of regular functionals can be constructed using the Stilties integral:

$$W[\phi] = \int_a^b \beta[x, \phi(x)]\, d\lambda(x),$$

$$W[\phi] = \int_a^b \alpha(\nu, \int_\nu^b \beta[x, \phi(x)]\, dx)\, d\lambda(\nu),$$

where $\beta : [-\tau, 0] \times \mathbf{R}^n \to \mathbf{R}$ are continuous functions, $\lambda : [-\tau, 0] \to \mathbf{R}$ is a function with bounded variation.

2.2 Singular functionals

Functionals of the form

$$V[\phi] = P(\phi(\zeta)), \quad \phi \in C[a, b], \tag{2.5}$$

$\Big(P(\cdot) : \mathbf{R} \to \mathbf{R}, \zeta \in (a, b)\Big)$ are called the *singular functionals*. For example:

$$V[\phi] = \phi\left(\frac{a+b}{2}\right) \tag{2.6}$$

is the singular functional.

Another example of the singular functional is the following

$$V[\phi] = \max_{a \le x \le b} \|\phi(x)\|. \tag{2.7}$$

3 Calculation of a functional along a line

3.1 Shift operators

Consider the procedure of calculation of the functional (1.4) along a continuous curve (function) [3] $\gamma(y)$, $y \in \mathbf{R}$.

[3] Without loss of generality we assume that the curve has infinite length (i.e. is defined on $\mathbf{R}$).

We have to consider the segment

$$\{\gamma(x+y)\,,\; a \le x \le b\}$$

of the function as the element of the space $C[a,b]$. Such representation is called the *segmentation principle* and allows one switch from the line $\gamma(y)$ to the set of continuous functions

$$T_y\gamma \equiv \{\gamma(x+y)\,,\; a \le x \le b\}\,. \tag{3.1}$$

The corresponding operator T_y is called the *shift operator.* For the shift operator we also use the notation $\gamma_y = T_y\gamma$.

Remark 3.1 One can introduce two types of shift operators:
a) r-shift operator (right shift operator)

$$T_y^+\gamma \equiv \{\gamma(x+y)\,,\; a \le x \le b\}\,; \tag{3.2}$$

b) l-shift operator (left shift operator)

$$T_y^-\gamma \equiv \{\gamma(x-y)\,,\; a \le x \le b\}\,. \tag{3.3}$$

□

Remark 3.2 We usually use the *right* shift operator T_y^+. So if there can be no misunderstanding we, as a rule, omit the sign "+", and write T_y instead of T_y^+. □

3.2 Superposition of a functional and a function

The function (1.4) calculated along the curve $\gamma(y)$ is the function

$$v(y) = V[T_y\gamma]\,,\; y \in \mathbf{R}\,, \tag{3.4}$$

i.e. the superposition of the functional V and the shift operator $T_y\gamma$.

3.3 Dini derivatives

For investigating properties of the functional (1.4) one can use *Dini derivatives* of the functional (1.4) along the function $\gamma(y)$:

$$D^{+}V[T_y\gamma] \equiv \limsup_{\Delta y \to 0} \frac{V[T_{y+\Delta y}\gamma] - V[T_y\gamma]}{\Delta y}, \tag{3.5}$$

$$D^{-}V[T_y\gamma] \equiv \liminf_{\Delta y \to 0} \frac{V[T_{y+\Delta y}\gamma] - V[T_y\gamma]}{\Delta y}. \tag{3.6}$$

If (3.4) is the differentiable function then

$$D^{+}V[T_y\gamma] = D^{-}V[T_y\gamma] = \dot{v}(y).$$

4 Discussion of two examples

4.1 Derivative of a function along a curve

Consider a differentiable function $g(x) : \mathbf{R} \to \mathbf{R}$ and a smooth curve[4] $x = \gamma(y)$, $y \in \mathbf{R}$. The derivative of the superposition

$$w(y) = g(\gamma(y)),\ y \in \mathbf{R}, \tag{4.1}$$

by the rule of differentiation of a superposition can be calculated as

$$\dot{w}(y) = \left.\frac{dg(x)}{dx}\right|_{x=\gamma(y)} \dot{\gamma}(y), \quad y \in \mathbf{R}. \tag{4.2}$$

In applications and concrete examples as a rule *elementary* functions or their combinations are usually used. The derivatives of these functions can be calculated *apriori*. In this case for calculating derivative of a function $g(x)$ *along the curve* $\gamma(y)$ one can formally substitute $\gamma(y)$ into beforehand calculated function (derivative) $\dfrac{dg(x)}{dx}$, and multiply the obtained expression by $\dot{\gamma}(y)$.

[4]which is also a smooth function.

4.2 Derivative of a functional along a curve

Calculation of the Dini derivative by the definition (see. paragraph 3.3) requires calculating the functional (1.4) along the curve γ, while for calculating derivative of the differentiable function $g(x)$ along the curve one can use the formula (4.2) which does not require calculating the function along the curve.

One can state the following *question*: can we calculate the derivative of the functional (1.4) along the curve γ by the analogy with the elementary function g? In other words consider the following

Question C: Is it possible define *apriori* a derivative ∂V of the functional (1.4) and obtain the derivative of the functional V along the function γ substituting γ into ∂V? □

Example 4.1 Let us calculate the derivative of the functional

$$V[\phi] = \int_a^b \alpha[\phi(x)]\, dx, \quad \phi \in C[a, b]\,, \tag{4.3}$$

along a curve $x = \gamma(y)$, $y \in \mathbf{R}$ (here $\alpha : \mathbf{R} \to \mathbf{R}$ is a continuous function). The functional (4.3) calculated along the curve γ is the function

$$v(y) \equiv V[T_y\gamma] = \int_a^b \alpha[\gamma(y + x)]\, dx\,, \tag{4.4}$$

The function (4.4) is differentiable and its derivative is

$$\frac{dV[T_y\gamma]}{dy} = \frac{d}{dy}\int_a^b \alpha[\gamma(y + x)]\, dx =$$

$$= \frac{d}{dy} \int\limits_{y+a}^{y+b} \alpha[\gamma(x)]\,dx = \alpha[\gamma(y+b)] - \alpha[\gamma(y+a)]\,. \tag{4.5}$$

Let us consider the functional

$$\partial V[\phi] = \alpha[\phi(b)] - \alpha[\phi(a)]\ ,\ \ \phi \in C[a,b]\ , \tag{4.6}$$

Then obviously

$$\dot{v}(y) = \partial V[T_y\gamma]\ . \tag{4.7}$$

Hence the derivative of the functional (4.3) along the curve γ coincides with the functional (4.6) calculated along the curve γ. □

Further the functional (4.7) is called the *invariant derivative* of the functional (4.3).

Example 4.2 Calculating the singular functional (2.6) along the curve γ. We obtain the superposition

$$v(y) \equiv V[T_y\gamma] = \gamma\left(y + \frac{a+b}{2}\right)\ . \tag{4.8}$$

The function $v(y)$ is differentiable if $\gamma(y)$ is the smooth (differentiable) curve. In this case

$$\frac{dV[T_y\gamma]}{dy} = \dot{v}(y) = \dot{\gamma}\left(y + \frac{a+b}{2}\right)\ . \tag{4.9}$$

Consider the functional

$$\partial V[\phi] = \dot{\phi}\left(\frac{a+b}{2}\right)\ ,\ \ \phi \in C[a,b]\ , \tag{4.10}$$

Then

$$\dot{v}(y) = \partial V[T_y\gamma]\ . \tag{4.11}$$

Hence the derivative of the functional (2.6) along the curve γ coincide with the functional (4.10) calculated along γ. □

5 The invariant derivative

5.1 The invariant derivative

In this section we give rigorous definition of the invariant derivative of the functional (1.4).

For $\phi \in C[a,b]$ and $\kappa > 0$ denote by

$$E_\kappa[\phi] \equiv \left\{ \Phi \in C[a-\kappa, b+\kappa] \;\middle|\; \Phi(x) = \phi(x),\, x \in [a,b] \right\}$$

the set of continuous extensions of the function ϕ, and by

$$[T_y\Phi](x) = \Phi(x+y)\,,\ \ a \leq x \leq b \in C[a,b]\,,$$

the *shift operator* of a function $\Phi \in E_\kappa[\phi]$ (i.e. contraction of the function Φ on the interval $[a+y, b+y]$). At that $T_\Phi \in C[a,b], y \in [-\kappa, \kappa]$.

Definition 5.1 *A functional* $V : C[a,b] \to \mathbf{R}$ *has at a point* $\phi \in C[a,b]$ *the invariant derivative* $\partial V[\phi]$*, if for any* $\Phi \in E_\kappa[\phi]$ *the corresponding function*

$$v_\Phi(y) = V[T_y\Phi]\,,\ \ y \in [-\kappa, \kappa], \tag{5.1}$$

has at the zero finite derivative $\dot{v}_\Phi(0)$ *invariant with respect to functions* $\Phi \in E_\kappa[\phi]$*, i.e. the value* $\dot{v}_\Phi(0)$ *of this derivative is the same for all* $\Phi \in E_\kappa[\phi]$*. In this case*

$$\partial V[\phi] = \dot{v}_\Phi(0). \tag{5.2}$$

□

Note that the invariant derivative does not depend on the value $\kappa > 0$ because it is defined as the limit.

If a functional V has at a point $\phi \in C[a,b]$ the invariant derivative, then we say that V is *invariantly differentiable* (*i-differentiable*) at the point ϕ. Operation of calculating the invariant derivative is called *i-differentiation* (invariant differentiation).

Example 5.1 Consider the linear on $C[a,b]$ functional

$$V[\phi] = \int_a^b \phi(x)\,dx\ , \tag{5.3}$$

and fix arbitrary $\phi \in C[a,b]$ and $\Phi \in E_\kappa[\phi]$. The corresponding function $v_\Phi(y) = V[T_y\Phi]$ has the form

$$v_\Phi(y) \equiv \int_a^b \Phi(y+x)\,dx = \int_{y+a}^{y+b} \Phi(x)\,dx\ ,$$

Its derivative at the point $y = 0$ has the form

$$\left.\frac{dv_\Phi(y)}{dy}\right|_{y=0} = \left(\frac{d}{dy}\int_{y+a}^{y+b} \Phi(x)\,dx\right)_{y=0} =$$

$$= \Phi(b) - \Phi(a) = \phi(b) - \phi(a)$$

and is invariant with respect to $\Phi \in E_\kappa[\phi]$. Thus the invariant derivative of the functional (5.3) is

$$\partial V[\phi] = \phi(b) - \phi(a)\ . \tag{5.4}$$

□

This example shows that the invariant derivative differs from the Frechet derivative (the Gateaux derivative) because the Frechet derivative of a linear continuous functional coincides with the functional.

5.2 The invariant derivative in the class $B[a,b]$

Continuity of a function $\phi \in C[a,b]$ and as the consequence of continuity of $\Phi \in E_\kappa[\phi]$ can be insufficient for differentiability of the function $v_\Phi(y)$ (for instance in the example 4.2, if γ is not the smooth curve). Therefore in this section we introduce the notion of the *invariant derivative in a class of functions* $B[a,b]$, where $B[a,b]$ is a space of sufficiently smooth functions, for example $C^1[a,b]$, $C^\infty[a,b]$, etc.

For $\phi \in B[a,b]$, $\kappa > 0$ denote

$$\hat{E}_\kappa[\phi] \equiv \left\{ \Phi : [a-\kappa, b+\kappa] \to \mathbf{R} \,\middle|\, \begin{array}{r} T_y\Phi \in B[a,b] \\ y \in [-\kappa,\kappa];\ T_0\Phi = \phi \end{array} \right\}.$$

Definition 5.2 *A functional*

$$V[\cdot] : B[a,b] \to \mathbf{R} \tag{5.5}$$

has at a point $\phi \in B[a,b]$ *the invariant derivative* $\partial V[\phi]$ *in the class* $B[a,b]$, *if for any* $\Phi \in \hat{E}_\kappa[\phi]$ *the function*

$$v_\Phi(y) = V[T_y\Phi],\ y \in [-\kappa,\kappa], \tag{5.6}$$

has at the point $y = 0$ *the derivative* $\dot{v}_\Phi(0)$ *invariant with respect to functions* $\Phi \in \hat{E}_\kappa[\phi]$. □

The *invariant derivative in the class* $B[a,b]$ is also called B*-invariant derivative*, for example, C^1-invariant derivative in case of $B[a,b] = C^1[a,b]$.

For $B[a,b] = C^m[a,b]$ the set $\hat{E}_\kappa[\phi]$ is denoted by $E^{(m)}_\kappa[\phi]$.

5.3 Examples

Example 5.2 We already noted in the example 4.1 that the derivative of the functional (2.1) has along the curves the invariance property. Let us calculate the invariant derivative of this functional by the definition 5.1; We assume that $\alpha : [a,b] \to \mathbf{R}$ is a continuous function.

Fix arbitrary functions $\phi \in C[a,b]$ and $\Phi \in E_\kappa[\phi]$. For the functional (2.1) the corresponding function $v_\Phi(y) = V[T_y\Phi]$ has the form

$$v_\Phi(y) \equiv \int_a^b \alpha[\Phi(y+x)]\,dx = \int_{y+a}^{y+b} \alpha[\Phi(x)]\,dx\,,$$

and its derivative

$$\left.\frac{dv_\Phi(y)}{dy}\right|_{y=0} = \left(\frac{d}{dy}\int\limits_{y+a}^{y+b} \alpha[\Phi(x)]\,dx\right)_{y=0} =$$

$$= \alpha[\Phi(b)] - \alpha[\Phi(a)] = \alpha[\phi(b)] - \alpha[\phi(a)]$$

is invariant with respect to $\Phi \in E_\kappa[\phi]$. Therefore the functional

$$\partial V[\phi] = \alpha[\phi(b)] - \alpha[\phi(a)] \tag{5.7}$$

is the invariant derivative of the functional (2.1). □

Example 5.3 Let in the functional

$$V[\phi] = \int\limits_a^b \alpha\left[\int\limits_\xi^b \beta[\phi(x)]\,dx\right] d\xi \tag{5.8}$$

$\alpha : \mathbf{R} \to \mathbf{R}$ is a continuous differentiable function, $\beta : \mathbf{R}^n \to \mathbf{R}$ is a continuous function.

To calculate the invariant derivative of the functional V at a point $\phi \in C[a,b]$ let us fix a function $\Phi \in E_\kappa[\phi]$ and consider

$$v_\Phi(y) \equiv \int\limits_a^b \alpha\,\Big[\int\limits_\xi^b \beta[\Phi(y+x)]\,dx\,\Big]\,d\xi =$$

$$= \int\limits_a^b \alpha\,\Big[\int\limits_{\xi+y}^y \beta[\Phi(x)]\,dx\,\Big]\,d\xi\,.$$

Calculate the derivative

$$\left.\frac{dv_\Phi(y)}{dy}\right|_{y=0} = \frac{d}{dy}\Big(\int\limits_a^b \alpha\,\Big[\int\limits_{\xi+y}^y \beta[\Phi(x)]ds\,\Big]\,d\xi\,\Big)_{y=+0} =$$

$$= \beta[\Phi(b)]\int\limits_a^b \dot{\alpha}\,\Big[\int\limits_\xi^b \beta[\Phi(x)]\,dx\,\Big]\,d\xi -$$

$$-\int_{-\tau}^{0} \dot{\alpha}\,\Big[\int_{\xi}^{b} \beta[\Phi(x)]\,dx\Big]\;\beta[\Phi(\xi)]\,d\xi =$$

$$= \beta[\Phi(b)] \int_{a}^{b} \dot{\alpha}\,\Big[\int_{\xi}^{b} \beta[\Phi(x)]\,dx\,\Big]\;d\xi -$$

$$-\int_{-\tau}^{0} \dot{\alpha}\,\Big[\int_{\xi}^{b} \beta[\Phi(x)]\,dx\Big]\;d\,\Big[\int_{\xi}^{b} \beta[\Phi(x)]\,dx\,\Big] =$$

$$= \beta[\Phi(b)] \int_{a}^{b} \dot{\alpha}\,\Big[\int_{\xi}^{b} \beta[\Phi(x)]\,dx\,\Big]\;d\xi +$$

$$+\int_{a}^{b} d\,\Big[\alpha\,\Big[\int_{\xi}^{b} \beta[\Phi(x)]\,dx\,\Big]\Big] =$$

$$= \beta[\Phi(b)] \int_{a}^{b} \dot{\alpha}\,\Big[\int_{\xi}^{b} \beta[\Phi(x)]\,dx\,\Big]\;d\xi +$$

$$+\alpha[0] - \alpha\,\Big[\int_{a}^{b} \beta[\Phi(x)]\,dx\,\Big]\;.$$

Taking into account that $\Phi(x) = \phi(x) \quad a \leq x \leq b$, obtain

$$\partial V[\phi] = \beta[\phi(b)] \int_{a}^{b} \dot{\alpha}\,\Big[\int_{\xi}^{b} \beta[\phi(x)]\,dx\,\Big]\;d\xi +$$

$$+\alpha(0) - \alpha\,\Big[\int_{a}^{b} \beta[\phi(x)]\,dx\,\Big]\;.$$

□

Example 5.4 Consider the functional

$$V[\phi] = \int_{a}^{b} \omega[x, \phi(x)]\,dx\;, \tag{5.9}$$

Where $\omega : [a,b] \times \mathbf{R} \to \mathbf{R}$ is a continuous differentiable function.

For $\phi \in C[a,b]$ and $\Phi \in E_\kappa[\phi]$ the corresponding function $v_\Phi(y)$ has the form

$$v_\Phi(y) \equiv \int_a^b \omega[x, \Phi(y+x)]\, dx = \int_{a+y}^{b+y} \omega[x-y, \Phi(x)]\, dx\,,$$

then

$$\left.\frac{dv_\Phi(y)}{dy}\right|_{y=0} = \frac{d}{dy}\Big(\int_{a+y}^{b+y} \omega[x-y, \Phi(x)]\, dx\Big)_{y=0} =$$

$$= \omega[b, \Phi(b)] - \omega[a, \Phi(a)] - \int_a^b \frac{\partial \omega[x, \Phi(x)]}{\partial x}\, dx\,,$$

where $\dfrac{\partial \omega[\cdot,\cdot]}{\partial x}$ is the partial derivative with respect to x.
$\Phi(x) = \phi(x)$ for $a \le x \le b$, then the functional

$$\partial V[\phi] = \omega[b, \phi(b)] - \omega[a, \phi(a)] - \int_a^b \frac{\partial \omega[x, \phi(x)]}{\partial x}\, dx$$

is the invariant derivative of the functional V. □

6 Properties of the invariant derivative

6.1 Principles of calculating invariant derivatives

Consider rules and formulas for calculating invariant derivatives without the application of definitions. Note that for the invariant derivative analogs of basic rules finite dimensional calculus is valid.

Theorem 6.1 *Let functional $V[\cdot], W[\cdot] : B[a,b] \to \mathbf{R}$ have at a point $\phi \in B[a,b]$ the invariant derivatives $\partial V[\phi]$ and*

$\partial W[\phi]$, then the sum and product of these functionals are invariant differentiable at the point ϕ and the following formulas are valid:

$$\partial\Big(V[\phi]+W[\phi]\Big)=\partial V[\phi]+\partial W[\phi]\,, \tag{6.1}$$

$$\partial\Big(V[\phi]\cdot W[\phi]\Big)=\partial V[\phi]\cdot W[\phi]+V[\phi]\cdot\partial W[\phi]\,. \tag{6.2}$$

Moreover, if $W[\phi]\neq 0$, then

$$\partial\left(\frac{V[\phi]}{W[\phi]}\right)=\frac{\partial V[\phi]\cdot W[\phi]-V[\phi]\cdot\partial W[\phi]}{W^2[\phi]}\,. \tag{6.3}$$

Proof. Fix an arbitrary function $\Phi\in\hat{E}_\kappa[\phi]$ and consider the functions

$$v_\Phi(y)\equiv V[T_y\Phi]\,, \tag{6.4}$$

$$w_\Phi(y)\equiv W[T_y\Phi]\,, \tag{6.5}$$

$y\in[-\kappa,\kappa]$. The sum of this function has the following derivative at zero

$$\frac{d}{dy}\Big(v_\Phi(y)+w_\Phi(y)\Big)\Big|_{y=0}=\dot{v}_\Phi(0)+\dot{w}_\Phi(0)=\partial V[\phi]+\partial W[\phi].$$

This does not depend on a specific form of Φ. Therefore the sum of the functionals V and W is invariantly differentiable at the point ϕ and the formula (6.1) is valid.

In a similar way, using the properties of the functions (6.4) and (6.5), one can derive the formulas (6.2) and (6.3). □

Remark 6.1 One can verify that the invariant derivative of a *constant functional* is equal to the zero. □

Remark 6.2 From (6.2) it follows that

$$\partial\Big(c\cdot V[\phi]\Big)=c\cdot\partial V[\phi]\,,$$

if a functional $V[\cdot]:B[a,b]\to\mathbf{R}$ is invariantly differentiable at a point $\phi\in B[a,b]$ and c is a constant. □

The invariant derivative of a superposition of a *function* and a *functional* can be calculated by the following rule.

Theorem 6.2 *If a functional $V : B[a,b] \to \mathbf{R}$ is invariantly differentiable at a point $\phi^* \in B[a,b]$ and a function $g(x) : \mathbf{R} \to \mathbf{R}$ is differentiable at the point $x^* = V[\phi^*]$ then the functional $G[\phi] = g(V[\phi])$ is invariantly differentiable at the point ϕ^* and*

$$\partial G[\phi^*] = \frac{d\,g(V[\phi^*])}{dx}\,\partial V[\phi^*]\,. \tag{6.6}$$

□

For example, if a functional V is invariantly differentiable at a point $\phi \in B[a,b]$, then

$$\partial\Big(c \cdot V[\phi]\Big) = c \cdot \partial V[\phi]\,,$$

$$\partial\Big(V^k[\phi]\Big) = kV^{k-1}[\phi] \cdot \partial V[\phi] \quad (k - const)\,,$$

$$\partial\Big(a^{V[\phi]}\Big) = a^{V[\phi]} \cdot \ln a \cdot \partial V[\phi]\,,$$

$$\partial\Big(e^{V[\phi]}\Big) = e^{V[\phi]}\,\partial V[\phi]\,,$$

$$\partial\Big(\log_a V[\phi]\Big) = \frac{\partial V[\phi]}{V[\phi]\,\ln a}\,,$$

$$\partial\Big(\ln V[\phi]\Big) = \frac{\partial V[\phi]}{V[\phi]}\,,$$

$$\partial\Big(\sin V[\phi]\Big) = \cos V[\phi] \cdot \partial V[\phi]\,,$$

$$\partial\Big(\cos V[\phi]\Big) = -\sin V[\phi] \cdot \partial V[\phi]\,,$$

$$\partial\Big(\arcsin V[\phi]\Big) = \frac{\partial V[\phi]}{\sqrt{1 - V^2[\phi]}},$$

$$\partial\Big(\arctan V[\phi]\Big) = \frac{\partial V[\phi]}{1 + V^2[\phi]}.$$

6.2 The invariant differentiability and invariant continuity

Consider the notion of the *differential* for functionals that have an invariant derivative

Definition 6.1 *Let the functional (5.5) have at a point $\phi \in B[a,b]$ the invariant derivative $\partial V[\phi]$ in a class $B[a,b]$. The product of this derivative and the difference $\Delta y \in \mathbf{R}$ $\big(|\Delta y| < \kappa\big)$*

$$dV = \partial V[\phi]\,\Delta y$$

is called an i-differential. □

It should be emphasized that i-differentials are the principle linear parts of the variation of the functional V along the curves that pass through ϕ. Moreover, the principle linear part does not depend on a specific form of curve $\Phi \in \hat{E}_\kappa[\phi]$! That is

$$\Delta V \equiv V[T_{b+\Delta y}\Phi] - V[\phi] = \partial V[\phi]\Delta y + O_\Phi(\Delta y) =$$

$$= dV + O_\Phi(\Delta y),$$

where the function $O_\Phi(\cdot) : [-\kappa, \kappa] \to \mathbf{R}$ depends on $\Phi \in \hat{E}_\kappa[\phi]$, and $\lim\limits_{\Delta y \to \infty} \dfrac{O_\Phi(\Delta y)}{\Delta y} = 0$.

Consider the notion of *invariant continuity*

Definition 6.2 *The functional (5.5) is B-invariantly continuous at a point $\phi \in B[a,b]$ if for any $\Phi \in \hat{E}_\kappa[\phi]$ the function $v_\Phi(y)$ (5.6) is continuous at the zero and the limit value $v_\Phi(0)$ is invariant with respect to Φ.* □

Invariant continuity and invariant differentiability of the functional (5.5) are defined by the appropriate properties of the function $\psi_\Phi(\xi)$, therefore, similar to the differentiability property of finite dimensional functions the following proposition is valid.

Theorem 6.3 *In order for the functional (5.5) to be invariant differentiable at a point $\phi \in B[a,b]$ it is necessary that at the point it had the invariant derivative $\partial_y V[\phi]$ and sufficient that the invariant derivative be B-invariant continuous at the point ϕ.* □

6.3 High order invariant derivatives

Consider for the functional (5.5) calculation of high order invariant derivatives at a point $\phi \in B[a,b]$. For all $\Phi \in \hat{E}_\kappa[\phi]$, $y \in (-\kappa,\kappa)$ let the functional (5.5) have at the point $T_y\Phi \in B[a,b]$ B-invariant derivative ∂V (Therefore the functional ∂V is defined on the whole set $\hat{E}_\kappa[\phi]$).

If the functional $\partial V : \hat{E}_\kappa[\phi] \to \mathbf{R}$ also has the B-invariant derivative at the point ϕ, then this B-invariant derivative is called a *B-invariant derivative of the second order* of the functional (5.5)and is denoted by $\partial^{(2)}V[\phi]$. If at the point ϕ there exists a m-th B-invariant derivative of the functional (5.5), then it is denoted by $\partial^{(m)}V[\phi]$.

Example 6.1 If $\phi \in C^m[a,b]$ then the singular functional (2.6) has C^m- invariant derivative up to the order m and

$$\partial^{(i)}V[\phi] = \phi^{(i)}\left(\frac{a+b}{2}\right) , \quad i = 1,\dots,m, \tag{6.7}$$

where $\phi^{(i)}$ is the i-th derivative of the function ϕ. □

Remark 6.3 Note that regular functionals (2.1) – (2.4) have invariant derivatives of the second order if their integrands have compact supports in (a,b). □

6.4 Series expansion

Let a functional

$$V : C^{\infty}[a,b] \to \mathbf{R} \tag{6.8}$$

has C^{∞}-invariant derivatives of all orders. Then it can be expanded into series along a function $\Phi \in \hat{E}_{\kappa}^{\infty}[\phi]$, $\kappa > 0$,

$$V[T_y\Phi] - V[\phi] = \sum_{i=1}^{\infty} \frac{\partial^{(i)} V[\phi]}{i!} \Delta y\,,$$

where $\partial^{(i)} V$ is the invariant derivative of i-th order

$$\Delta y = \begin{cases} y - b, & \text{if } \ y > b\,, \\ a - y, & \text{if } \ y < a\,. \end{cases}$$

7 Several variables

7.1 Notation

Consider the constructions of the invariant derivative in case of a functional

$$V[\cdot] : C(\bar{\Omega}) \to \mathbf{R}\,, \tag{7.1}$$

where $\bar{\Omega}$ is the closed subset of $\mathbf{R}^n$, $C(\bar{\Omega})$ is the set of continuous functions $\phi : \bar{\Omega} \to \mathbf{R}$. $\bar{\Omega}_{\kappa}$ denotes the closed κ–neighbourhood of the set $\bar{\Omega}$.

For $\phi \in C(\bar{\Omega})$ and $\kappa > 0$ consider the set

$$E_{\kappa}[\phi] \equiv \left\{ \Phi \in C(\bar{\Omega}_{\kappa}) \;\middle|\; \Phi(x) = \phi(x),\, x \in \bar{\Omega} \right\} \tag{7.2}$$

of continuous extensions of the function ϕ.

7.2 Shift operator

For $\phi \in C(\bar{\Omega})$, $\kappa > 0$ and $\Phi \in E_{\kappa}[\phi]$ the shift operator is defined by the formula

$$T_y\Phi = \left\{\Phi(x+y), x \in \bar{\Omega}\right\} \in C(\bar{\Omega}), \tag{7.3}$$

$y \in \mathbf{R}^n$, $\|y\| \leq \kappa$.

7.3 Partial invariant derivative

Definition 7.1 *The functional (7.1) has at a point $\phi \in C(\bar{\Omega})$ the partial invariant derivative $\partial_k V[\phi]$ with respect to x_k, if for every $\Phi \in E_\kappa[\phi]$ the function*

$$v_\Phi(y) = V[T_y\Phi], \ \|y\| \leq \kappa, \tag{7.4}$$

has at the zero $y = 0$ the partial derivative $\dfrac{\partial v_\Phi(0)}{\partial y_k}$ invariant with respect to $\Phi \in E_\kappa[\phi]$, i.e. the value $\dfrac{\partial v_\Phi(0)}{\partial y_k}$ is the same for all $\Phi \in E_\kappa[\phi]$. In this case

$$\partial_k V[\phi] = \frac{\partial v_\Phi(0)}{\partial y_k}. \tag{7.5}$$

□

Remark 7.1 Similar to section 5.2. one can introduce the notion of the *invariant derivative in a class* of sufficiently smooth functions $B(\bar{\Omega}) \subset C(\bar{\Omega})$. □

8 Generalized derivatives of nonlinear functionals

8.1 Introduction

Theories of distributions and generalized functions [8, 32, 34] were developed for investigating various linear differential equation classes in functional spaces of functions and functionals possessing generalized derivatives. However the unsolved problem of defining associative and commutative product of distributions [8, 32] is an obstacle for the study of nonlinear differential equations in the distribution space.

In this section we develop the theory of *generalized derivatives of nonlinear* on $D(a, b)$ functionals (nonlinear distributions) and the approach to the notion of generalized solutions of nonlinear differential equations[5].

A notion of the generalized derivative of nonlinear on $D(a, b)$ functionals is based on the constructions applied for introducing the invariant derivative. Nevertheless the term "generalized derivative" is used and in the nonlinear case, because for linear continuous functionals the introduced generalized derivative coincides with the classic generalized derivative of the distribution theory.

This section presents an approach to the notion of generalized solutions of nonlinear differential equations closed to [15]. However it is based on the new notion of the generalized derivative.

We study nonlinear differential equations in a space SD of infinite differentiable in the generalized sense nonlinear on $D(a, b)$ functionals, which are called generalized distributions ($D'(a, b) \in SD$).

The convenience of our approach is due to the construction of a generalized calculus of nonlinear functionals that is similar to the construction of the classic finite dimensional calculus.

In section 8.8 we prove the existence of a *primitive* for any element of the space SD. The proof is based on the existence in D of a *basis on shift* analogous, in some sense, to a linear independent basis in a vector space.

A notion of generalized solutions of nonlinear differential equations is introduced in the section 8.9. The theorem about solvability in SD linear differential equations with nonlinear time variable coefficients is proved.

---

[5]Beginning with the works [4, 5, 15] in the framework of distribution theory *nonlinear* theory of multiplication distributions are developed.

8.2 Distributions (generalized functions)

For the sake of simplicity we consider the one-dimensional case [6]. Terms distribution and generalized functions are used as the synonyms.

Hereinafter $D = D(a,b)$ is the space of infinite differentiable functions with compact supports in $(a,b) \subset \mathbf{R}$ (the space of test functions). $D' = D'(a,b)$ is the set of linear continuous on $D = D(a,b)$ functionals (the space of distributions (generalized functions), and (f,ϕ) denotes the value of a distribution $f \in D'$ at a test function $\phi \in D$.

By definition (see, for example, [32]) the generalized derivative of a distribution $f \in D'$ is a distribution $f' \in D'$ such that,

$$(f',\phi) = -(f,\phi'), \ \phi \in D. \tag{8.1}$$

Emphasize the important fact that the generalized derivative can be also defined on the basis of the (left) shift operator,

$$[T_y\phi](x) \equiv \phi(x-y), \ a \leq x \leq b. \tag{8.2}$$

(see for example [8]) as the limit,

$$(f',\phi) = \lim_{y\to 0}\left[\frac{(f,T_y\phi)-(f,\phi)}{y}\right]. \tag{8.3}$$

Below in section T_y is the *left translation operator.*

For the correct application of the shift operator consider that each basic function $\phi \in D(a,b)$ is extended as the zero from (a,b) on the whole $\mathbf{R}$. Therefore the shift $T_y\phi$ is defined for all $\phi \in D(a,b)$ and $y \in \mathbf{R}$. Moreover, it is an element of more wide space $D(\mathbf{R})$.

If a function ϕ has a compact support then due to the openness of the interval (a,b) for sufficiently small y the inclusion $T_y\phi \in D(a,b)$ is valid.

[6] All results by analogy can be extended on the case of an open set $\Omega \subset \mathbf{R}^n$.

8.3 Generalized derivatives of nonlinear distributions

Basic definitions

For a *nonlinear* functional

$$u : D(a,b) \to \mathbf{R} \tag{8.4}$$

denotes by $\prec u, \phi \succ$ as its value at $\phi \in D$.

Definition 8.1 *A (nonlinear) functional $u' : D \to \mathbf{R}$ is called the generalized derivative of the (nonlinear) functional (8.4) if for every $\phi \in D$*

$$\prec u', \phi \succ = \lim_{y \to 0} \frac{\prec u, T_y\phi \succ - \prec u, \phi \succ}{y}. \tag{8.5}$$

□

Remark 8.1 For distributions[7] $f \in D'$ the generalized derivative (in the sense of the definition 8.1) coincides due to (8.3) with the classic generalized derivative in the sense of the distribution theory. □

Remark 8.2 Note, although the invariant derivative was defined *locally* (at a point), for the convenience of exposition we define the generalized derivative of functionals *globally* (similarly to the classic linear distributions) as a functional on the whole D. □

Remark 8.3 By u'' or $u^{(2)}$ we define the generalized derivative of the first generalized derivative $u' : D \to \mathbf{R}$. In general the notation $u^{(j)}$ is used for the j-th generalized derivative (if it exists) of a nonlinear functional $u : D \to \mathbf{R}$. □

[7] linear continuous functionals

Generalized derivative as the invariant derivative

The generalized derivative of a nonlinear functional (8.4) can be defined as C^∞-invariant derivative.

For a function $\phi \in D(a,b)$ and a constant $\kappa > 0$ consider the set

$$\hat{E}_\kappa[\phi] \equiv \left\{ \Phi \in C^\infty[a-\kappa, b+\kappa] \;\middle|\; \begin{array}{c} T_y\Phi \in D(a,b) \text{ for} \\ y \in [-\kappa,\kappa]; T_0\Phi = \phi \end{array} \right\}.$$

The following definition is equivalent to the definition 8.1.

Definition 8.2 *The functional (8.4) has at an element $\phi \in D$ a generalized derivative, denoting $\prec u', \phi \succ$, if for any $\Phi \in \hat{E}_\kappa[\phi]$ the corresponding function*

$$v_\Phi(y) = \prec u, T_y\Phi \succ\,, \;\; y \in [-\kappa,\kappa], \tag{8.6}$$

has at the point $y = 0$ the derivative $\dot{v}_\Phi(0)$, invariant with respect to $\Phi \in \hat{E}_\kappa[\phi]$, i.e. the value $\dot{v}_\Phi(0)$ is the same for all $\Phi \in \hat{E}_\kappa[\phi]$.

Note the generalized derivative does not depend on a specific value of $\kappa > 0$, because it is defined as the limit. The following valid proposition:

Theorem 8.1 $\Big(\forall \phi \in D(a,b)\Big)\; \Big(\forall \kappa > 0\Big)\; \Big(\forall \Phi \in \hat{E}_\kappa[\phi]\Big)$ $\Big(\exists \kappa_\Phi \in (0,\kappa]\Big)$ *Such that*

$$\Phi(x) = \begin{cases} 0, & x \in [a-\kappa_\Phi, a]\,, \\ \phi(x), & x \in (a,b)\,, \\ 0, & x \in [b, b+\kappa_\Phi]\,. \end{cases}$$

□

From the theorem 8.1 $\Phi \in D(a-\kappa_\Phi, b+\kappa_\Phi)$ follows. Therefore, to introduce the generalized derivative by the formula (8.5), one can use just one function $\Phi \in E_\kappa[\phi]$ (equal to the zero outside the interval (a,b)).

8.4 Properties of generalized derivatives

One can prove that, if nonlinear functionals $u, v : D \to \mathbf{R}$ have at an element $\phi \in D$ generalized derivatives u' and v', then their *sum* $(u+v)$ has at the point ϕ the invariant derivative and the following equality is valid

$$\prec (u+v)', \phi \succ = \prec u', \phi \succ + \prec v', \phi \succ . \qquad (8.7)$$

The generalized derivative of the superposition of a *function* and a *nonlinear functional* can be calculated by the following rule

Theorem 8.2 *If a nonlinear functional $u : D \to \mathbf{R}$ has at a point $\phi^* \in D$ the generalized derivative and a function $g(x) : \mathbf{R} \to \mathbf{R}$ is differentiable at the point $x^* = \prec u, \phi^* \succ$, then the nonlinear functional $G : D \to \mathbf{R}$ acting by the rule*

$$\prec G, \phi^* \succ = g(\prec u, \phi \succ) , \quad \phi \in D ,$$

has at the point ϕ^* the generalized derivative and

$$\prec G', \phi^* \succ = \left. \frac{d\, g(x)}{dx} \right|_{x = \prec u, \phi^* \succ} \prec u', \phi^* \succ . \qquad (8.8)$$

□

For example if a nonlinear functional $u : D \to \mathbf{R}$ has the generalized derivative at a point $\phi \in D$ then

$$\prec (c\, u)', \phi \succ = c \prec u', \phi \succ \quad (c - const) ,$$

$$\prec \left(u^k\right)', \phi \succ = k \prec u^{k-1}, \phi \succ \prec u', \phi \succ \quad (k - const) ,$$

$$\left(a^{\prec u, \phi \succ}\right)' = a^{\prec u, \phi \succ} \prec u', \phi \succ \ln a ,$$

$$\left(e^{\prec u, \phi \succ}\right)' = e^{\prec u, \phi \succ} \prec u', \phi \succ ,$$

$$\left(\log_a \prec u, \phi \succ\right)' = \frac{\prec u', \phi \succ}{\prec u, \phi \succ \ln a} ,$$

$$\left(\ln \prec u, \phi \succ\right)' = \frac{\prec u', \phi \succ}{\prec u, \phi \succ},$$

$$\left(\sin \prec u, \phi \succ\right)' = (\cos \prec u, \phi \succ) \prec u', \phi \succ,$$

$$\left(\cos \prec u, \phi \succ\right) = -(\sin \prec u, \phi \succ) \prec u', \phi \succ.$$

8.5 Generalized derivative (multidimensional case)

In this section we consider the invariant derivative of nonlinear distributions

$$u : D(\Omega) \to \mathbf{R}, \tag{8.9}$$

where Ω is an open set of $\mathbf{R}^n$, $D(\Omega)$ is the space of infinite differentiable functions $\phi : \Omega \to \mathbf{R}$ with compact supports in Ω. The space $D(\Omega)$ can be considered as the subset of $D(\mathbf{R}^n)$. The value of the functional (8.9) at an element $\phi \in D(\Omega)$ we denote by $\prec u, \phi \succ$.

For $\phi \in D(\Omega)$ and $\kappa > 0$ the *shift operator* is defined by the rule

$$T_y \phi = \left\{\phi(x+y), x \in \Omega\right\} \in D(\mathbf{R}^n), \quad y \in \mathbf{R}^n, \ \|y\| \le \kappa.$$

Note that $T_y \phi \in D(\Omega)$ for $y \in \mathbf{R}^n$ with sufficiently small norm is similar to the definition 8.2 and one can introduce the notion of the *partial generalized derivative* of the functional (8.9).

Definition 8.3 *A nonlinear functional (8.9) has at an element $\phi \in D(\Omega)$ the partial generalized derivative (with respect to x_k), denoting by $\prec \partial_k u, \phi \succ$, if the function*

$$v(y) \equiv \prec u, T_y \phi \succ \tag{8.10}$$

has at the point $y = 0$ the partial derivative $\dfrac{\partial v(0)}{\partial y_k}$ with *respect to y_k. In this case*

$$\prec\partial_k u, \phi\succ = \frac{\partial v(0)}{\partial y_k}.$$

□

8.6 The space SD of nonlinear distributions

The introduction of the new derivative allows us to consider a class of spaces $S^m D$ which consist of nonlinear on D functionals, having continuous (in topology D) generalized derivatives up to the m-th order.

In this section we study the solvability of nonlinear differential equations in the space $SD \equiv S^\infty D$ of infinitely smooth (in the generalized sense) nonlinears on D functionals.

Elements of the space SD are called *generalized distributions* or *nonlinear distributions.* Obviously, the distribution space D' is the subset of SD.

Another examples of generalized distributions are functionals of the form $u = a[f]$ $(f \in D', a \in C^\infty(\mathbf{R}))$, Acting by the rule $\prec u, \phi\succ = a[(f, \phi)]$, $\phi \in D$. It easy to verify that $\prec u', \phi\succ = a'[(f, \phi)](f', \phi)$, etc.

The set of analytical on D functionals [15] is the subspace in SD.

Convergence in SD

Definition 8.4 *Directivity $(u_\alpha)_{\alpha\in A} \subset SD$ converges to $u \in SD$, if for all $\phi \in D$ and $j \in N$ the sequence $(\prec u_\alpha^{(j)}, \phi\succ)_{\alpha\in A}$ converges to $\prec u^{(j)}, \phi\succ$.* □

Multiplication in SD

The space SD is the algebra with respect to the operation of *pointwise* multiplication (denoting by $\odot$) defined by the rule: for v, $w \in SD$

$$\prec v \odot w, \phi \succ = \prec v, \phi \succ \prec w, \phi \succ \ , \quad \phi \in D \ . \tag{8.11}$$

Obviously this product is associative, commutative and the Leibniz differentiation principle is valid:

$$(v \odot w)' = v' \odot w + v \odot w' \ . \tag{8.12}$$

Because in the space of the direct product $SD \times SD$ a directivity converges to a point from $(SD \times SD)$ if and only if its projection into the arbitrary coordinate space converges to the projection of this point, then applying the mathematical induction method, one can prove that the mapping $\{v, w\} \to v \odot w$ from $SD \times SD$ to SD is continuous, hence SD is topological algebra.

Note, if $u, v \in SD$ and $\prec v, \phi \succ \neq 0$, then

$$\prec \left(\frac{u}{v}\right)', \phi \succ = \frac{\prec u' \odot v, \phi \succ - \prec u \odot v', \phi \succ}{\prec v^2, \phi \succ} \ . \tag{8.13}$$

Consider several formulas involving multiplication and differentiation operations:

$$(\arcsin u)' = \frac{u'}{\sqrt{1-u^2}} \ , \qquad (\arctan u)' = \frac{u'}{1+u^2} \ .$$

8.7 Basis on shift

In the space D there is a basis which will play a principal role in further constructions.

For a subset A of D denote

$$dis(A) = \{T_y \phi \ \big| \ \phi \in A, y \in \mathbf{R}\} \bigcap D(a, b) \ .$$

Definition 8.5 *A set $B \subset D$ is called*

1) *Independent on shift (translation independent), if for any $\psi \in B$*

$$\psi \notin dis(B \setminus \{\psi\}) \ ,$$

2) Basis on a shift (translation basis), if it is independent on shift and $disB = D$.

□

Theorem 8.3 *There exists a translation basis in* D.

Proof. Denote by M the set of all independent on shift subsets of D (obviously $M \neq \emptyset$).

The inclusion operations determine the partial order on M. Every chain $C \subset M$ has the superior bound, namely unification of all the subsets of D which are the same time elements of C.

By the Zorn lemma the set M contains the maximal element B. Let us show that B is the *basis on shift* in D. Let $Y = dis(B)$. Then Y is the subset of D and, moreover, $Y = D$, because otherwise $B \bigcup \{z\}$, $z \in D$, $z \notin Y$, would be the *independent on the shift* set, containing B as a subset, that contradicts the maximum of B. □

8.8 Primitive

One of the fundamental notions of every differential equation theory is the notion of the primitive, which allows one find solutions in specific cases and construct a general theory.

Definition 8.6 *A generalized distribution* $\int u \in SD$ *is called the primitive of a generalized distribution* $u \in SD$ *if* $\left(\int u\right)' = u$. □

In this section we prove the existence of the primitive for every generalized distribution.

The primitive is defined with accuracy up to a constant.

Constants in SD are defined in the sense of the following

Definition 8.7 *A generalized distribution q is called the functional constant (in SD), if* $\prec q', \phi \succ = 0$ *for all* $\phi \in D$.
□

It is clear that constants of D' may also be constants in SD. Additionally in SD there are nonlinear constants, that are, for example identical[8] on D functionals. Another class of (nonlinear) constants present functionals, $q \in SD$, acting by the rule:

$$\prec q, \phi \succ = \int_a^b \phi^m(x)\, dx \;, \quad \phi \in D \;, m \in N \;. \tag{8.14}$$

Now we prove the existence of the primitive for an arbitrary generalized distribution $u \in SD$.

Fix a translation basis B. The value of the primitive $\int u$ at $\phi \in D$ is calculated in the following way. For ϕ there is a unique $\psi \in B$ and $y \in \mathbf{R}$ such that

$$\phi = T_y \psi \,, \tag{8.15}$$

Then we set

$$\prec \int u, \phi \succ = \int_0^y \prec u, T_x\psi \succ dx + \prec q, \phi \succ \,, \tag{8.16}$$

where q are functional constants (in the sense of definition 8.7).

It is easy to verify that the functional, determined by relations (8.16) and (8.15) belongs to SD.

Note that our construction of the primitive depends on the choice of the translation basis B. However primitives of a generalized distribution $u \in SD$, constructed on a different translation bases, differ only on a constant in SD. Thus the following proposition is valid.

[8] constant on the whole space

Theorem 8.4 *Each generalized distribution $u \in SD$ has a unique accuracy up to a constant the primitive $\int u \in SD$.*

□

From this theorem follows that there exists in SD the solution of the differential equation

$$u' = g \quad (g \in SD) \tag{8.17}$$

of the form $u = \int g + q$, where $\int g$ is a primitive and q is a functional constant.

Let us show that the finding of a primitive is the *linear* operation.

Theorem 8.5 *Let $u, v \in SD$ and $k \in \mathbf{R}$. Then*

$$\int k\,u = k \int u\,, \tag{8.18}$$

$$\int [u + v] = \int u + \int v\,, \tag{8.19}$$

$$\int [u - v] = \int u - \int v\,. \tag{8.20}$$

Proof. To prove the formulas (8.18) and (8.19), sufficiently differentiate the right hand sides and see that we obtain Integrands of the left hand sides:

$$\left(\int k\,u\right)' = k\left(\int u\right)' = k\,u\,,$$

$$\left(\int u + \int v\right)' = \left(\int u\right)' + \left(\int v\right)' = u + v\,.$$

Hence formulas (8.18) and (8.19) are valid.

The formula (8.20) follows from (8.18) and (8.19). □

The following proposition is valid.

Theorem 8.6 *Let $u \in SD$, $k \in \mathbf{R}$, $k \neq -1$. Then*

$$\int u^k\, u' = \frac{u^{k+1}}{k+1} + q\,, \tag{8.21}$$

where q is a functional constant.

Proof. For $u \in SD$ and $k \in \mathbf{R}$, $k \neq -1$, the following differentiation formula is valid

$$\left(\frac{u^{k+1}}{k+1}\right)' = u^k\, u'\,. \tag{8.22}$$

Integrating both sides of (8.22) we obtain (8.21). □

8.9 Generalized solutions of nonlinear differential equations

At first consider the simplest differential equation in SD:

$$u' = f \odot u \qquad (f \in D') \tag{8.23}$$

By the direct substitution one can verify that the *general solution* of the equation (8.23) is the functional $u = q \odot e^{\int f}$ acting by the rule

$$\prec u, \phi \succ = \prec q, \phi \succ e^{(\int f, \phi)}\,, \qquad \phi \in D, \tag{8.24}$$

where $\displaystyle\int f$ is a primitive of f and, q is a functional constant (because $e^x \in C^\infty(R)$ and $\displaystyle\int f \in D'$, hence $u \in SD$).

For example the equation

$$u' = \delta(x) \odot u$$

has the solution $e^{\theta(x)}$, where $\theta(x)$ is the Heaviside step function.

Let us define the general notion of the solution of a differential equation in the space SD. Let $F[u;\zeta] : SD \times D \to SD$. Consider the equation

$$u' = F[u;\zeta]\,, \qquad \zeta \in D\,, \tag{8.25}$$

with respect to the unknown functional $u \in SD$.

Let B be a translation basis in D and q is a constant in SD.

Definition 8.8 *The generalized solution of the equation (8.25) with the initial condition $q \in SD$ is a functional (generalized solution) $u \in SD$ which for every $\phi \in D$ satisfies the equation*

$$\prec u', T_y\phi \succ = \prec F[u; T_y\phi], T_y\phi \succ \quad \textit{for} \quad y \in \mathbf{R}, \tag{8.26}$$

and the (initial) condition

$$\prec u, \psi \succ = \prec q, \psi \succ, \qquad \psi \in B. \tag{8.27}$$

□

Example 8.1 Consider the equation

$$u' = 3u^{1/3} \odot \delta(x) \odot \theta(x) .$$

By the direct substitution one can verify that a particular generalized solution of this equation is $u = \theta^3$. □

Example 8.2 To find the general solution of the nonlinear equation

$$u' = \delta(x) \odot \mid u \mid . \tag{8.28}$$

Let us fix a translation basis B and a constant q. For every $\psi \in B$ associate the function

$$a_\psi(y) = \prec q, \psi \succ e^{sign \prec q, \psi \succ \int_0^y (\delta(x), T_x\psi) dx},$$

which is the solution on $\mathbf{R}$ the Cauchy problem

$$\begin{cases} \dfrac{d}{dy} a(y) = (\delta(x), T_y\psi) \cdot \mid a(y) \mid, \\ a(0) = \prec q, \psi \succ . \end{cases} \tag{8.29}$$

It is clear that system (8.29) corresponds to system (8.26)–(8.27) (for the equation (8.28)). For every $\phi \in D$ there is the unique $\psi \in B$ and $y \in \mathbf{R}$ such that $\phi = T_y\psi$. We set $\prec u, \phi \succ = a_\psi(y)$. □

Remark 8.4 Thus our approach reduces the solvability problem of the equation (8.25) in the space SD to the solution of the corresponding (in the our case ordinary) differential equations (8.26) [with the initial conditions (8.27)] along shifts of the elements of the translation basis B. Emphasize, if coefficients of the equation (8.25) are singular (for example $\delta(x)$), coefficients of the equation (8.26) can still be smooth (in the classic sense) with respect to y. □

8.10 Linear differential equations with variables coefficients

In this section we prove the theorem about solvability in SD the linear differential equation with variables coefficients

$$Pu \equiv \sum_{j=0}^{m} g_j \odot u^{(j)} = h, \tag{8.30}$$

where $h,\ g_j \in D'$; $u^{(j)}$ is the generalized derivative of j-th order ($j = 0, \dots, m$) The desired generalized distribution u. The *generalized solution* of the equation (8.30) is a functional $u \in SD$, which for every $\phi \in D$ satisfy the differential equation

$$\prec Pu,\, T_y\phi \succ \, = \, \prec h,\, T_y\phi \succ, \qquad y \in \mathbf{R}.$$

For single valued solvability, the equation (8.30) necessary by means of some constants q_j ($j = 0, \dots, m-1$) assign the initial conditions at some translation basis B.

Theorem 8.7 *Unique in the space SD there is the generalized solution of the equation (8.30), satisfying the conditions:*

$$u^{(j)} = q_j, \quad (j = 0, \dots, m-1).$$

Proof. Let us describe the general scheme of calculating the value of the solution u at an element $\phi \in D$. By virtue of the properties of the translation basis B there are the unique $\psi \in B$ and $y \in \mathbf{R}$ such that

$$\phi = T_y \psi \,. \tag{8.31}$$

From the theory of ordinary linear differential equations follows the existence of the exclusive function $a_\psi(\cdot) \in C^\infty(R)$, satisfying the equation

$$\sum_{j=0}^{m} \prec g_j, T_x\psi \succ \left(\frac{d^{(j)}}{dx^j} a_\psi(x) \right) = \prec h, T_x\psi \succ \,, \quad x \in \mathbf{R}\backslash\{0\}, \tag{8.32}$$

and the conditions

$$\frac{d^{(j)}}{dx^j} a_\psi(0) = \prec q_j, \psi \succ \,, \quad j = 0, \ldots, m-1.$$

Now, by definition we set $\prec u, \phi \succ = a_\psi(y)$.

Infinite smoothness (in the generalized sense) of the solution u follows from the construction technique, and the continuity of the solution and all its generalized derivatives on D follows from the continuous dependence on the initial conditions of the solution of linear differential equation (8.32) □

9 Functionals on $Q[-\tau, 0)$

Fundamental notion which is used for describing and investigating functional differential equations is the notion of a functional

$$V[y(\cdot)] : Q[-\tau, 0) \to \mathbb{R}. \tag{9.1}$$

The goal of this section is to describe the basic classes of *elementary functionals*, which are used in FDE theory. Elementary functionals are analogs of *elementary functions* [9] that compose the basis of the finite-dimensional calculus.

[9] polynomials, trigonometric functions, rational functions, etc.

In the this section we develop the direction of i-smooth calculus of functionals which is applied in FDE theory.

Consider an example of a *linear* on $Q[-\tau, 0)$ functional.

Example 9.1 Let $k[\cdot] : [-\tau, 0] \to \mathbb{R}^n$ be a continuous function, $a \in \mathbb{R}^n$, $\tau_* \in [-\tau, 0)$. Then

$$V[y(\cdot)] = \int_{-\tau}^{0} k'(s)y(s)ds + a'\, y(-\tau_*) \tag{9.2}$$

is the linear functional on $Q[-\tau, 0)$. □

Going after V. Volterra (see., for example, [38], p. 59) in the functional (9.2) the part, represented by an integral, is called the *regular* component, and the second part is called the *discrete (singular) component.*

Further we will use the same terminology and for *nonlinear* functionals:

- *regular* functionals are presented in integral forms over $Q[-\tau, 0)$,
- *singular* functionals depend on values of a function $y(\cdot) \in Q[-\tau, 0)$ at a finite number of points of the interval $Q[-\tau, 0)$.

Functionals of these two types are usually used in practice [10].

Consider two examples of widely used functionals.

Example 9.2 (regular nonlinear functional) Let $\alpha[\cdot] : \mathbb{R}^n \to \mathbb{R}$ be a continuous function, then

$$V[y(\cdot)] = \int_{-\tau}^{0} \alpha[y(s)]ds \tag{9.3}$$

is the *regular* functional on $Q[-\tau, 0)$. □

[10] Rigorous mathematical notions of *regular* and *singular* functionals can be defined in terms of *absolutely continuous* and *discrete* measures. However it is not the goal of the section.

Example 9.3 (singular nonlinear functional) Let $P[\cdot] : \mathbb{R}^n \to \mathbb{R}$ be a continuous function, then

$$V[y(\cdot)] = P[y(-\tau)] \tag{9.4}$$

is the *singular* functional on $Q[-\tau, 0)$. □

A discussion of the rigorous notions of *regular* and *singular (discrete)* functionals in nonlinear cases is not the goal of the present section, so below we present only examples of specific functionals of these forms.

9.1 Regular functionals

Consider examples of regular functionals.

Example 9.4 Let $\beta[\cdot,\cdot] : [-\tau, 0] \times \mathbb{R}^n \to \mathbb{R}$ be a continuous function, then

$$V[y(\cdot)] = \int_{-\tau}^{0} \beta[s, y(s)]ds \tag{9.5}$$

is the *regular* functional on $Q[-\tau, 0)$. □

Example 9.5 Let $\gamma[\cdot,\cdot] : [-\tau, 0] \times \mathbb{R} \to \mathbb{R}$ be a continuous function, then

$$V[y(\cdot)] = \int_{-\tau}^{0} \gamma[\xi, \int_{\xi}^{0} \beta[s, y(s)]ds]d\xi \tag{9.6}$$

is the *regular* functional on $Q[-\tau, 0)$. □

Example 9.6 Let $\omega[\cdot,\cdot;\cdot,\cdot] : [-\tau, 0] \times \mathbb{R}^n \times [-\tau, 0] \times \mathbb{R}^n \to \mathbb{R}$ be a continuous function, then

$$V[y(\cdot)] = \int_{-\tau}^{0}\int_{-\tau}^{0} \omega[s, y(s); \xi, y(\xi)]dsd\xi \tag{9.7}$$

is the *regular* functional on $Q[-\tau, 0)$. □

If α, β, γ are continuous functions, then integrals in the right parts of the described above functionals are defined for every function $y(\cdot) \in Q[-\tau, 0)$.

Example 9.7 A special case of functional (9.7) is the quadratic functional (regular homogeneous functional of the second degree, [38], p. 63)

$$V[y(\cdot)] = \int_{-\tau}^{0} \int_{-\tau}^{0} y'(s)\gamma[s,\nu]\, y(\nu) ds d\nu, \tag{9.8}$$

where $\gamma[s,\nu]$ $(s,\nu \in [-\tau,0])$ is a continuous $n \times n$ matrix with continuous elements. □

In general case one can construct regular functionals, described by m-multiple integrals. However in practice integrals of order $m > 3$ are applied in rare cases.

9.2 Singular functionals

A singular functional was described in example 9.3. In general, singular functionals can be defined in a similar manner.

Example 9.8 (singular functional) Let $P^*[\cdot,\ldots,\cdot] : \mathbb{R}^n \times \ldots \times \mathbb{R}^n \to \mathbb{R}$ be a continuous function and $\tau_1,\ldots,\tau_m \in (0,\tau]$, Then

$$V[y(\cdot)] = P^*[y(-\tau_1),\ldots,y(-\tau_m)] \tag{9.9}$$

is the *singular* functional on $Q[-\tau, 0)$. □

9.3 Specific functionals

There are functionals that do not belong to regular or singular types. Such functionals will be called *specific* or *special* functionals. For example, the specific be the following functional

$$V[y(\cdot)] = \sup_{-\tau \leq s \leq 0} \|y(s)\| \, ; \tag{9.10}$$

$$V[y(\cdot)] = \phi(y([s])) \, , \quad -\tau \leq s \leq 0 \, , \tag{9.11}$$

where $\phi(\cdot) : \mathbb{R}^n \to \mathbb{R}$, $[s]$ is the greatest integer in s.

9.4 Support of a functional

In this section we discuss a notion of a support of a functional

$$V[y(\cdot)] : Q[-\tau, 0) \to \mathbb{R}. \tag{9.12}$$

Consider a subset $\sigma \subseteq [-\tau, 0)$ and denote by $y_\sigma(\cdot) = \{y(s), s \in \sigma\}$ the restriction[11] of a function $y(\cdot) \in Q[-\tau, 0)$ on the set σ. Introduce the set $Q_\sigma = \{y_\sigma(\cdot) : y(\cdot) \in Q[-\tau, 0)\}$ of restrictions of functions $y(\cdot)$ on σ.

Definition 9.1 *A set $\sigma \subseteq [-\tau, 0)$ is called the support of the functional (9.12), if there exists a functional*

$$U[y_\sigma(\cdot)] : Q_\sigma \to \mathbb{R} \tag{9.13}$$

such that

$$V[y(\cdot)] = U[y_\sigma(\cdot)]$$

for all $y(\cdot) \in Q[-\tau, 0)$.

Example 9.9 Consider regular functional

$$V[y(\cdot)] = \int_{-\tau}^{-\varepsilon} \beta[y(s)] ds, \tag{9.14}$$

where $\varepsilon \in (0, \tau)$, $\beta[\cdot] : \mathbb{R}^n \to \mathbb{R}$ is a continuous function. Obviously the set $\sigma = [-\tau, -\varepsilon]$ is the support of (regular) functional (9.12). □

Example 9.10 Singular functional (9.4) has the point support $\sigma = \{-\tau\}$. The support of the functional (9.9) consists of a finite set $\sigma = \{-\tau_1, \ldots, -\tau_m\}$. □

[11] constriction

Definition 9.2 *The functional (9.12) has a support $\sigma \subseteq [-\tau, 0)$, separated from zero, if there exists a constant $\Delta \in (0, \tau)$ such that $\sigma \bigcap [-\Delta, 0) = \emptyset$.*

Note that in all examples of this subsection the supports of the functionals were separated from zero.

10 Functionals on $\mathbb{R} \times \mathbb{R}^n \times Q[-\tau, 0)$

In the general case the right hand sides of the functional differential equations depends on variables t, x $y(\cdot)$, i.e. have the form

$$V[t, x, y(\cdot)] : \mathbb{R} \times \mathbb{R}^n \times Q[-\tau, 0) \to \mathbb{R}. \tag{10.1}$$

In concrete cases functionals (10.1) are constructed as combinations of:

— finite-dimensional functions $g(t, x)$;
— regular (integral) functionals;
— singular functionals;
— specific functionals.

One of the most simple and natural forms of such a functionals is the form

$$V[t, x, y(\cdot)] = g(t, x) + W[y(\cdot)], \tag{10.2}$$

where $g(t, x) : \mathbb{R} \times \mathbb{R}^n \to \mathbb{R}$ and $W[y(\cdot)] : Q[-\tau, 0) \to \mathbb{R}$.

Functionals (10.1) are called *regular* (*singular*), if for any $(t_*, x^*) \in \mathbb{R} \times \mathbb{R}^n$ functional $V[t_*, x^*, y(\cdot)] : Q[-\tau, 0) \to \mathbb{R}$ is a regular (singular) functional.

10.1 Regular functionals

Consider examples of regular functionals of the form (10.2).

Example 10.1 Functional

$$V[t, x, y(\cdot)] = g(t, x) + \int_{-\tau}^{0} \beta[s, y(s)]ds \tag{10.3}$$

is the regular functional if $\beta[\cdot,\cdot] : [-\tau, 0] \times \mathbb{R}^n \to \mathbb{R}$ is a continuous function. $\square$

Example 10.2 Functional

$$V[t, x, y(\cdot)] = g(t, x) + \int_{-\tau}^{0} \gamma[\xi, \int_{\xi}^{0} \beta[s, y(s)]ds]d\xi \quad (10.4)$$

is the regular functional if $\beta[\cdot,\cdot]$ and $\gamma[\cdot,\cdot] : [-\tau, 0] \times \mathbb{R} \to \mathbb{R}$ are continuous functions. $\square$

Example 10.3 Functional

$$V[t, x, y(\cdot)] = g(t, x) + \int_{-\tau}^{0}\int_{-\tau}^{0} \omega[s, y(s); \xi, y(\xi)]dsd\xi. \quad (10.5)$$

is the regular functional if $\omega[\cdot,\cdot;\cdot,\cdot] : [-\tau, 0] \times \mathbb{R}^n \times [-\tau, 0] \times \mathbb{R}^n \to \mathbb{R}$ is a continuous function. $\square$

Functionals like (10.1) can be of a more complex form

Example 10.4 Functional

$$V[t, x, y(\cdot)] = \int_{-\tau}^{0} \eta[t, s, x, y(s)]ds \quad (10.6)$$

is the regular functional if $\eta[\cdot,\cdot,\cdot,\cdot] : \mathbb{R} \times \mathbb{R} \times \mathbb{R}^n \times \mathbb{R}^n \to \mathbb{R}$ is a continuous function. $\square$

Example 10.5 Functional

$$V[t, y(\cdot)] = \int_{-\tau^*(t)}^{0} \mu[s, y(s)]ds \quad (10.7)$$

is the regular functional if $\mu[\cdot,\cdot] : [-\tau, 0] \times \mathbb{R}^n \to \mathbb{R}$ is a continuous function, $\tau^*(\cdot) : \mathbb{R} \to (0, \tau]$. $\square$

10.2 Singular functionals

Example 10.6 Let (10.2) the component W be a singular functional, for example (9.4), then

$$V[t,x,y(\cdot)] = g(t,x) + P[y(-\tau)] \tag{10.8}$$

be the singular on $\mathbb{R} \times \mathbb{R}^n \times Q[-\tau, 0)$ functional. □

Singular functionals (10.1) can have more complicated dependence on t and x.

Example 10.7 Singular functional

$$V[t,y(\cdot)] = P[y(-\tau^*(t))], \tag{10.9}$$

$\tau^*(\cdot) : \mathbb{R} \to (0,\tau]$. □

Example 10.8 Singular functional

$$V[t,x,y(\cdot)] = P[y(-\tau^{\#}(t,x))], \tag{10.10}$$

$\tau^{\#}(\cdot) : \mathbb{R} \times \mathbb{R}^n \to (0,\tau]$. □

10.3 Volterra functionals

For describing and investigating systems with unbounded delays the following classes of functional are used:

Example 10.9 Volterra functionals

$$V[t,x,y(\cdot)] = g(t,x) + \int_{-t}^{0} \sigma(t,t+s)P[y(s)]ds\,, \tag{10.11}$$

$$V[t,x,y(\cdot)] = g(t,x) + \int_{-t}^{0}\int_{\nu}^{0} P[y(s)]dsd\nu\,, \tag{10.12}$$

$$V[t,x,y(\cdot)] = g(t,x) + \int_{-t}^{0}\int_{-t}^{-\nu} \sigma(t+s+\nu,t+s)P[y(s)]dsd\nu\,, \tag{10.13}$$

$\sigma(\cdot,\cdot) : \mathbb{R} \times \mathbb{R} \to \mathbb{R}$ and $P[\cdot] : \mathbb{R}^n \to \mathbb{R}$ are continuous functions. These are regular functionals. □

Remark 10.1 Regular and singular functionals, described in this chapter, and all those functionals that can be obtained from these by using addition, subtraction, multiplication, division, composition with elementary functions and taking of inverse functions are called *elementary functionals.*

10.4 Support of a functional

Let us introduce the notion of a support for functionals of the form (10.1).

Definition 10.1 *a) A set $\sigma_{t_*,x^*} \subseteq [-\tau, 0)$ is called the support of functional (10.1), at a point $(t_*, x^*) \in \mathbb{R} \times \mathbb{R}^n$, if there exists a functional (9.13) such that*

$$V[t_*, x^*, y(\cdot)] = U[y_\sigma(\cdot)]$$

for all $y(\cdot) \in Q[-\tau, 0)$.
b) A set $\sigma \subseteq [-\tau, 0)$ is called the (uniform) support of functional (10.1), if for every $(t, x) \in \mathbb{R} \times \mathbb{R}^n$ there exists a functional

$$U_{(t,x)}[y_\sigma(\cdot)] : Q_\sigma \to \mathbb{R}$$

such that

$$V[t, x, y(\cdot)] = U_{(t,x)}[y_\sigma(\cdot)]$$

for all $y(\cdot) \in Q[-\tau, 0)$.

11 The invariant derivative

If in a functional

$$V[t, x, y(\cdot)] : \mathbb{R} \times \mathbb{R}^n \times Q[-\tau, 0) \to \mathbb{R} \tag{11.1}$$

the functional variable $y(\cdot)$ is fixed, then the functional becomes a finite-dimensional function of variables t and x. In this case we say that functional (11.1) has partial derivative $\frac{\partial V}{\partial t}$ with respect to t and gradient $\frac{\partial V}{\partial x}$ with respect to x, if the corresponding finite-dimensional function has the corresponding partial derivatives.

For example, the functional (10.2) has partial derivatives

$$\frac{\partial V[t, x, y(\cdot)]}{\partial t} = \frac{\partial g(t, x)}{\partial t}, \quad \frac{\partial V[t, x, y(\cdot)]}{\partial x} = \frac{\partial g(t, x)}{\partial x},$$

if the function $g(t, x)$ has the corresponding derivatives, because the (functional) term $W[y(\cdot)]$ does not depend on t and x, hence its partial derivatives with respect to t and x are equal to zero.

The main goal of this section is to introduce the notion of the *invariant derivative* of the functional (11.1) with respect to the functional variable $y(\cdot)$.

11.1 Invariant derivative of a functional

Consider auxiliary sets and constructions.

For $x \in \mathbb{R}^n$, $y(\cdot) \in Q[-\tau, 0)$ and $\beta > 0$ denote by $E_\beta[x, y(\cdot)]$ the set of all continuous extentions [12] of the pair $\{x, y(\cdot)\}$ on the interval $[0, \beta]$, i.e. $E_\beta[x, y(\cdot)]$ is the set of functions $Y(s) : [-\tau, \beta] \to \mathbb{R}^n$ such that:

1) $Y(0) = x$,

2) $Y(s) = y(s), \quad -\tau \leq s < 0$,

3) $Y(s)$ is continuous on $[0, \beta]$.

Further $E[x, y(\cdot)] = \bigcup\limits_{\beta > 0} E_\beta[x, y(\cdot)]$, i.e for $Y(\cdot) \in E[x, y(\cdot)]$ there exists $\beta > 0$ such that $Y(\cdot) \in E_\beta[x, y(\cdot)]$.

Consider the function

$$\Psi_Y(\eta, z, \xi) = V[t + \eta, x + z, y_\xi(\cdot)], \tag{11.2}$$

[12] continuations

where $\eta > 0$, $z \in \mathbb{R}^n$, $\xi \in [0, \beta]$, $y_\xi(\cdot) = \{Y(\xi+s), -\tau \le s < 0\} \in Q[-\tau, 0)$. Note, the function (11.2) and the interval $[0, \beta]$ depend on the choice of $Y(\cdot) \in E[x, y(\cdot)]$.

Definition 11.1 *The invariant derivative $\partial_y V$ of a functional $V[t, x, y(\cdot)]$ is the right-hand partial derivative of the function $\Psi_Y(\eta, z, \xi)$ with respect to ξ at zero, if this partial derivative exists and is invariant*[13] *with respect to $Y(\cdot) \in E[x, y(\cdot)]$:*

$$\partial_y V = \left. \frac{\partial \Psi_Y(\eta, z, \xi)}{\partial \xi} \right|_{\eta=0, z=0, \xi=+0}.$$

Note that existence of the invariant derivative depends on local properties of the function (11.2) in the right neighborhood of zero, so in definition11.1 the set $E[x, y(\cdot)]$ can be replaced by $E_\beta[x, y(\cdot)]$ for some $\beta > 0$.

In terms of the function (11.2) one can define partial derivatives of functional (11.1) with respect to t and x.

Definition 11.2 *The partial derivative $\dfrac{\partial V}{\partial t}$ of the functional $V[t, x, y(\cdot)]$ is the right-hand partial derivative of the function $\Psi_Y(\eta, z, \xi)$ with respect to η at zero:*

$$\frac{\partial V}{\partial t} = \left. \frac{\partial \Psi_Y(\eta, z, \xi)}{\partial \eta} \right|_{\eta=+0, z=0, \xi=+0}.$$

Definition 11.3 *The partial derivative $\dfrac{\partial V}{\partial x_i}$, $i \in \{1, \ldots, n\}$, of the functional $V[t, x, y(\cdot)]$ is the right-hand partial derivative of the function $\Psi_Y(\eta, z, \xi)$ with respect to z_i at zero:*

$$\frac{\partial V}{\partial x_i} = \left. \frac{\partial \Psi_Y(\eta, z, \xi)}{\partial z_i} \right|_{\eta=0, z=0, \xi=+0}.$$

[13] has the same value

Definition 11.4 *The gradient* $\dfrac{\partial V}{\partial x}$ *of the functional* $V[t,x,y(\cdot)]$ *is the vector of partial derivatives* $\left(\frac{\partial V}{\partial x_1},\cdots,\frac{\partial V}{\partial x_n}\right)'$.

11.2 Examples

Example 11.1 Consider a functional

$$V[y(\cdot)] = \int_{-\tau}^{0} \alpha(y(s))ds \tag{11.3}$$

($\alpha(\cdot) : \mathbb{R}^n \to \mathbb{R}$ is a continuous function), defined on $Q[-\tau,0)$. Note that the functional (11.3) does not depend on x, nevertheless the invariant derivative is calculating at a point $h = \{x, y(\cdot)\} \in \mathbb{R}^n \times Q[-\tau,0)$, which includes x.

Let us fix an arbitrary function $Y(\cdot) \in E[x, y(\cdot)]$ and consider the function

$$\Psi_Y(\xi) = V[y_\xi(\cdot)] = \int_{-\tau}^{0} \alpha(Y(\xi+s))ds = \int_{-\tau+\xi}^{\xi} \alpha(Y(s))ds.$$

Applying the rule of differentiating integrals with variable limits and taking into account that $Y(0) = x$ and $Y(-\tau) = y(-\tau)$, we obtain

$$\frac{d\Psi_Y(0)}{d\xi} = \frac{d}{d\xi}\Big(\int_{-\tau+\xi}^{\xi} \alpha(Y(s))ds\Big)_{\xi=+0} =$$

$$= \alpha(Y(0)) - \alpha(Y(-\tau)) = \alpha(x) - \alpha(y(-\tau)).$$

Thus, the derivative $\dfrac{d\Psi_Y(0)}{d\xi} = \alpha(x) - \alpha(y(-\tau))$ is an invariant with respect to continuations $Y(\cdot) \in E[x, y(\cdot)]$ and depends only on the point $\{x, y(\cdot)\}$. Hence the functional (11.3) has at every point $h = \{x, y(\cdot)\} \in Q[-\tau,0)$ the invariant derivative $\partial_y V[x, y(\cdot)] = \alpha(x) - \alpha(y(-\tau))$. □

Emphasize once more, that thought the functional (11.3) depends only on $y(\cdot)$, we calculate its invariant derivative $\partial_y V[x, y(\cdot)]$ at a pair $\{x, y(\cdot)\} \in H$. That is for calculating invariant derivatives of regular functionals the boundary values of test functions play the essential role. By this reason, though the functional (11.3) is defined on $L_2[-\tau, 0)$[14] – it does not have invariant derivatives on functions from $L_2[-\tau, 0)$. The issue is that elements $y(\cdot) \in L_2[-\tau, 0)$ are classes of functions that differ on sets of zero measure. Hence for $y(\cdot) \in L_2[-\tau, 0)$ the value $\alpha[y(-\tau)]$ is not defined. However if a function $y(\cdot) \in L_2[-\tau, 0)$ is continuous to the right at $s = -\tau$, then for functional (11.3) one can calculate at the point $\{x, y(\cdot)\} \in \mathbb{R}^n \times L_2[-\tau, 0)$ the invariant derivative, which similar to the case of piece-wise continuous function $y(\cdot)]$ be equal to $\partial_y V = \alpha(x) - \alpha(y(-\tau))$.

Example 11.2 Consider a functional

$$V[t, x, y(\cdot)] = g(t, x) + \int_{-\tau}^{0} \alpha(y(s))ds\,, \tag{11.4}$$

where $g(\cdot, \cdot) : \mathbb{R} \times \mathbb{R}^n \to \mathbb{R}$ is a continuous differentiable function and $\alpha(\cdot) : \mathbb{R}^n \to \mathbb{R}$ is a continuous function.

Construct the function

$$\Psi_Y(\eta, z, \xi) = V[t + \eta, x + z, Y_\xi(\cdot)] =$$

$$= g(t + \eta, x + z) + \int_{-\tau}^{0} \alpha(Y(\xi + s))ds =$$

$$= g(t + \eta, x + z) + \int_{-\tau+\xi}^{\xi} \alpha(Y(s))ds\,.$$

[14]if the integral in (11.3) understands as Lebesgue integral

One can easily calculate partial derivatives with respect to t and x:

$$\frac{\partial V[t,x,y(\cdot)]}{\partial t} = \frac{\partial g(t,x)}{\partial t}, \qquad \frac{\partial V[t,x,y(\cdot)]}{\partial x} = \frac{\partial g(t,x)}{\partial x}.$$

The invariant derivative is equal to

$$\partial_y V[t,x,y(\cdot)] = \alpha(x) - \alpha(y(-\tau)). \qquad (11.5)$$

Note, the invariant derivative of the function $g(t,x)$ is equal to zero, because the function does not depend on $y(\cdot)$), so formula (11.5) directly follows from Example 11.1. □

Example 11.3 Consider the singular functional

$$V[y(\cdot)] = y(-\tau).$$

The corresponding function (11.2) has the form

$$\Psi_Y(\xi) = V[Y_\xi(\cdot)] = y(\xi - \tau) \quad \text{for} \quad \xi \in [0,\tau).$$

If the function $y(s)$, $-\tau \le s < 0$ is differentiable at the point $s = -\tau$, then the functional has the invariant derivative $\partial_y V = \dot{y}(-\tau)$. □

Note, for calculating the invariant derivative of *singular* functionals we do not use continuations $Y(\cdot) \in E[x,y(\cdot)]$ of the function $y(\cdot)$.

This is the specific feature of functionals with supports separated from zero.

Example 11.4 Let in the functional

$$V[t,x,y(\cdot)] = g(t,x) + \int_{-\tau}^{0} \alpha\Big(\int_{\nu}^{0} \beta[y(s)]ds\Big)d\nu \qquad (11.6)$$

$g(\cdot,\cdot) : \mathbb{R} \times \mathbb{R}^n \to \mathbb{R}$ and $\alpha(\cdot) : \mathbb{R} \to \mathbb{R}$ be continuous differentiable functions, and let $\beta[\cdot] : \mathbb{R}^n \to \mathbb{R}$ be a continuous function. The integral in the right part of (11.6) does not depend on t and x, hence

$$\frac{\partial V[t,x,y(\cdot)]}{\partial t}=\frac{\partial g(t,x)}{\partial t}\quad\text{and}\quad\frac{\partial V[t,x,y(\cdot)]}{\partial x}=\frac{\partial g(t,x)}{\partial x}.$$

To calculate the invariant derivative fix a continuation $Y(\cdot)\in E[x,y(\cdot)]$ and consider the function

$$\Psi_Y(\eta,z,\xi)=g(t+\eta,x+z)+\int_{-\tau}^{0}\alpha\Big(\int_{\nu}^{0}\beta[Y(\xi+s)]ds\Big)\,d\nu=$$

$$=g(t+\eta,x+z)+\int_{-\tau}^{0}\alpha\Big(\int_{\nu+\xi}^{\xi}\beta[Y(s)]ds\Big)\,d\nu\,.$$

One can calculate this by applying the theorem of differentiating under integral and formula of total derivative of a composition

$$\frac{d\Psi_Y(0,0,0)}{d\xi}=\frac{d}{d\xi}\Big(\int_{-\tau}^{0}\alpha\Big(\int_{\nu+\xi}^{\xi}\beta[Y(s)]ds\Big)\,d\nu\Big)_{\xi=+0}=$$

$$=\beta[Y(0)]\int_{-\tau}^{0}\dot\alpha\Big(\int_{\nu}^{0}\beta[Y(s)]ds\Big)\,d\nu-$$

$$-\int_{-\tau}^{0}\dot\alpha\Big(\int_{\nu}^{0}\beta[Y(s)]ds\,\beta[Y(\nu)]\Big)\,d\nu=$$

$$=\beta[Y(0)]\int_{-\tau}^{0}\dot\alpha\Big(\int_{\nu}^{0}\beta[Y(s)]ds\Big)\,d\nu+\int_{-\tau}^{0}d\Big[\alpha\Big(\int_{\nu}^{0}\beta[Y(s)]ds\Big)\Big]=$$

$$=\beta[Y(0)]\int_{-\tau}^{0}\dot\alpha\Big(\int_{\nu}^{0}\beta[Y(s)]ds\Big)\,d\nu+\alpha(0)-\alpha\Big(\int_{-\tau}^{0}\beta[Y(s)]ds\Big).$$

Taking into account that $Y(0) = x$ and $Y(s) = y(s)$ for $-\tau \leq s < 0$, obtain

$$\partial_y V[p] = \beta[x] \int_{-\tau}^{0} \dot{\alpha}\left(\int_{\nu}^{0} \beta[y(s)]ds\right) d\nu + \alpha(0) - \alpha\left(\int_{-\tau}^{0} \beta[y(s)]ds\right).$$

□

Example 11.5 Consider the functional

$$V[t, x, y(\cdot)] = \int_{-\tau}^{0} \omega[t, s, x, y(s)]\, ds\,, \qquad (11.7)$$

where $\omega[\cdot,\cdot,\cdot,\cdot] : \mathbb{R} \times [-\tau, 0] \times \mathbb{R}^n \times \mathbb{R}^n \to \mathbb{R}$ is a continuous differentiable function. The corresponding function (11.2) has the form

$$\Psi_Y(\eta, z, \xi) = \int_{-\tau}^{0} \omega[t+\eta, s, x+z, Y(\xi+s)]ds\,.$$

Here $p = \{t, x, y(\cdot)\} \in \mathbb{R} \times \mathbb{R}^n \times Q[-\tau, 0)$ and $Y(\cdot) \in E[x, y(\cdot)]$. Obviously

$$\frac{\partial V[p]}{\partial t} = \int_{-\tau}^{0} \frac{\partial \omega[t, s, x, y(s)]}{\partial t} ds\,,$$

$$\frac{\partial V[p]}{\partial x} = \int_{-\tau}^{0} \frac{\partial \omega[t, s, x, y(s)]}{\partial x} ds\,.$$

(Note, these partial derivatives can be obtained by directly differentiating the functional (11.7) with respect to t and x).

The function $\Psi_Y(\eta, z, \xi)$ can be presented in the form

$$\Psi_Y(\eta, z, \xi) = \int_{-\tau+\xi}^{\xi} \omega[t+\eta, s-\xi, x+z, Y(s)]ds\,,$$

Then

$$\frac{d\Psi_Y(0,0,0)}{d\xi} =$$

$$= \frac{d}{d\xi}\Big(\int_{-\tau+\xi}^{\xi} \omega[t+\eta, s-\xi, x+z, Y(s)]ds \Big)_{\xi=+0} =$$

$$= \omega[t,0,x,Y(0)] - \omega[t,-\tau,x,Y(-\tau)] -$$

$$- \int_{-\tau}^{0} \frac{\partial \omega[t,s,x,Y(s)]}{\partial s} ds \,,$$

where $\frac{\partial \omega}{\partial s}$ is the derivative with respect to the second variable. Taking into account that $Y(0) = x$ and $Y(s) = y(s)$ for $-\tau \leq s < 0$, obtain

$$\partial_y V[t,x,y(\cdot)] = \omega[t,0,x,x] - \omega[t,-\tau,x,y(-\tau)] -$$

$$- \int_{-\tau}^{0} \frac{\partial \omega[t,s,x,y(s)]}{\partial s} ds \,.$$

□

Example 11.6 Consider the functional

$$V[x,y(\cdot)] = \int_{-\tau}^{0} \int_{\nu}^{0} \omega[x,s,y(s)]dsd\nu \,, \qquad (11.8)$$

where $\omega[\cdot,\cdot,\cdot] : \mathbb{R}^n \times [-\tau,0] \times \mathbb{R}^n \to \mathbb{R}$ is a continuous differentiable function. For this functional the corresponding function (11.2) has the form

$$\psi_Y(z,\xi) = \int_{-\tau}^{0} \int_{\nu}^{0} \omega[x+z,s,Y(\xi+s)]dsd\nu$$

for $Y(\cdot) \in E[x, y(\cdot)]$.

Obviously

$$\frac{\partial V[p]}{\partial x} = \int_{-\tau}^{0} \int_{\nu}^{0} \frac{\partial \omega[x, s, y(s)]}{\partial x} ds d\nu .$$

Present the function $\psi_Y(z, \xi)$ in the form

$$\psi_Y(z, \xi) = \int_{-\tau}^{0} \int_{\nu+\xi}^{\xi} \omega[x + z, s - \xi, Y(s)] ds d\nu ,$$

then

$$\frac{d\psi_Y(0, 0)}{d\xi} =$$

$$= \frac{d}{d\xi} \left(\int_{-\tau}^{0} \int_{\nu+\xi}^{\xi} \omega[x + z, s - \xi, Y(s)] ds d\nu \right)_{\xi=+0} =$$

$$= \tau\, \omega[x, 0, Y(0)] - \int_{-\tau}^{0} \omega[x, s, Y(s)] ds -$$

$$- \int_{-\tau}^{0} \int_{\nu}^{0} \frac{\partial \omega[x, s, Y(s)]}{\partial s} ds d\nu ,$$

where $\frac{\partial \omega}{\partial s}$ – is the derivative with respect to the second variable. Taking into account that $Y(0) = x$ and $Y(s) = y(s)$ for $-\tau \leq s < 0$, obtain the following formula for the invariant derivative

$$\partial_y V[x, y(\cdot)] = \tau \omega[x, 0, x] - \int_{-\tau}^{0} \omega[x, s, y(s)] ds -$$

$$-\int_{-\tau}^{0}\int_{\nu}^{0}\frac{\partial\omega[x,s,y(s)]}{\partial s}dsd\nu\,.$$

□

Example 11.7 Consider the functional

$$V[y(\cdot)] = \int_{-\tau}^{0}\left[\left(\int_{\nu}^{0} y(s)ds\right)'\Gamma\left(\int_{\nu}^{0} y(s)ds\right)\right]d\nu\,, \quad (11.9)$$

where Γ is a symmetric $n\times n$ matrix. The corresponding function (11.2) has the form

$$\psi_Y(\xi) = \int_{-\tau}^{0}\left[\left(\int_{\nu}^{0} Y(s+\xi)ds\right)'\Gamma\left(\int_{\nu}^{0} Y(s+\xi)ds\right)\right]d\nu$$

for $Y(\cdot)\in E[x,y(\cdot)]$.

The function $\psi_Y(\xi)$ can be presented in The form

$$\psi_Y(\xi) = \int_{-\tau}^{0}\left[\left(\int_{\nu+\xi}^{\xi} Y(s)ds\right)'\Gamma\left(\int_{\nu+\xi}^{\xi} Y(s)ds\right)\right]d\nu\,,$$

Hence

$$\frac{d\psi_Y(0)}{d\xi} = \frac{d}{d\xi}\left(\int_{-\tau}^{0}\left[\left(\int_{\nu+\xi}^{\xi} Y(s)ds\right)'\Gamma\left(\int_{\nu+\xi}^{\xi} Y(s)ds\right)\right]d\nu\right)_{\xi=+0} =$$

$$= \int_{-\tau}^{0}\left[\Big(Y(0)-Y(\nu)\Big)'\Gamma\left(\int_{\nu}^{0} Y(s)ds\right)\right]d\nu +$$

$$+\int_{-\tau}^{0}\left[\left(\int_{\nu}^{0} Y(s)ds\right)'\Gamma\Big(Y(0)-Y(\nu)\Big)\right]d\nu\,.$$

Thus

$$\partial_y V[x,y(\cdot)] = \int_{-\tau}^{0} \left[\Big(x - y(\nu)\Big)' \Gamma \left(\int_{\nu}^{0} y(s)ds \right) \right] d\nu +$$

$$+ \int_{-\tau}^{0} \left[\left(\int_{\nu}^{0} y(s)ds \right)' \Gamma \Big(x - y(\nu)\Big) \right] d\nu =$$

$$= 2x'\Gamma \int_{-\tau}^{0} \int_{\nu}^{0} y(s)dsd\nu - 2 \int_{-\tau}^{0} y'(\nu)\Gamma \left(\int_{\nu}^{0} y(s)ds \right) d\nu =$$

$$= 2x'\Gamma \int_{-\tau}^{0} \int_{\nu}^{0} y(s)dsd\nu - 2 \int_{-\tau}^{0} \left(\int_{\nu}^{0} y(s)\, ds \right)' \Gamma y(\nu)d\nu =$$

$$= 2x'\Gamma \int_{-\tau}^{0} \int_{\nu}^{0} y(s)dsd\nu + 2 \int_{-\tau}^{0} \left(\int_{\nu}^{0} y(s)ds \right)' \Gamma d \left[\int_{\nu}^{0} y(s)ds \right] =$$

$$= 2x'\Gamma \int_{-\tau}^{0} \int_{\nu}^{0} y(s)dsd\nu + \int_{-\tau}^{0} d \left[\left(\int_{\nu}^{0} y(s)ds \right)' \Gamma \left(\int_{\nu}^{0} y(s)ds \right) \right] =$$

$$= 2x'\Gamma \int_{-\tau}^{0} \int_{\nu}^{0} y(s)dsd\nu - \left(\int_{-\tau}^{0} y(s)ds \right)' \Gamma \left(\int_{-\tau}^{0} y(s)ds \right).$$

□

Example 11.8 Consider calculation of the invariant derivative of the functional

$$V[x,y(\cdot)] = \int_{-\tau}^{0} \int_{-\tau}^{0} \gamma[x; s, y(s); u, y(u)]dsdu\,.$$

For a function $Y(\cdot) \in E[x,y(\cdot)]$ consider

$$\psi_Y(z,\xi) = \int_{-\tau}^{0} \int_{-\tau}^{0} \gamma[x+z; s, Y(s+\xi); u, Y(u+\xi)]dsdu\,,$$

then

$$\frac{d\psi_{Y}(0,0)}{d\xi} =$$

$$= \frac{d}{d\xi}\left(\int_{-\tau+\xi}^{\xi}\int_{-\tau+\xi}^{\xi} \gamma[x; s-\xi, Y(s); u-\xi, Y(u)]dsdu\right)_{\xi=+0} =$$

$$= \int_{-\tau}^{0} \gamma[x; 0, Y(0); u, Y(u)]du - \int_{-\tau}^{0} \gamma[x; -\tau, Y(-\tau); u, Y(u)]du +$$

$$+ \int_{-\tau}^{0} \gamma[x; s, Y(s); 0, Y(0)]ds - \int_{-\tau}^{0} \gamma[x; s, Y(s); -\tau, Y(-\tau)]ds -$$

$$- \int_{-\tau}^{0}\int_{-\tau}^{0} \frac{\partial\gamma[x; s, Y(s); u, Y(u)]}{\partial s} dsdu -$$

$$- \int_{-\tau}^{0}\int_{-\tau}^{0} \frac{\partial\gamma[x; s, Y(s); u, Y(u)]}{\partial u} dsdu\,.$$

Hence

$$\partial_y V[x, y(\cdot)] =$$

$$= \int_{-\tau}^{0} \gamma[x; 0, x; u, y(u)]du - \int_{-\tau}^{0} \gamma[x; -\tau, y(-\tau); u, y(u)]du +$$

$$+ \int_{-\tau}^{0} \gamma[x; s, y(s); 0, x]ds - \int_{-\tau}^{0} \gamma[x; s, y(s); -\tau, y(-\tau)]ds -$$

$$- \int_{-\tau}^{0}\int_{-\tau}^{0} \frac{\partial\gamma[x; s, y(s); u, y(u)]}{\partial s} dsdu -$$

$$- \int_{-\tau}^{0}\int_{-\tau}^{0} \frac{\partial\gamma[x; s, y(s); u, y(u)]}{\partial u} dsdu\,.$$

One can verify that

$$\frac{\partial V[x,y(\cdot)]}{\partial x} = \int\limits_{-\tau}^{0}\int\limits_{-\tau}^{0} \frac{\partial \gamma[x; s, y(s); u, y(u)]}{\partial x} ds du\,.$$

□

11.3 Invariant continuity and invariant differentiability

In the subsection we consider the notions of the *invariant continuity*(*i*-continuity) and the *invariant differentiability* (*i*-differentiability) of functionals.

Definition 11.5 *The functional (11.1) is invariantly continuous (i-continuous) at a point* $p = \{t, x, y(\cdot)\} \in \mathbb{R} \times \mathbb{R}^n \times Q[-\tau, 0)$, *if for any* $Y(\cdot) \in E[x, y(\cdot)]$ *the corresponding function (11.2) is continuous at the zero (continuity with respect to* η *and* ξ *from the right) and the limit value* $\Psi_Y(0)$ *is invariant with respect to* $Y(\cdot)$.

Definition 11.6 *The functional (11.1) is invariantly differentiable (i–differentiable) at a point* $p = \{t, x, y(\cdot)\} \in \mathbb{R} \times \mathbb{R}^n \times Q[-\tau, 0)$, *if at this point exist finite derivatives* $\dfrac{\partial V[p]}{\partial t}$, $\dfrac{\partial V[p]}{\partial x}$, $\partial_y V[p]$ *and for any* $Y(\cdot) \in E[x, y(\cdot)]$ *the following equality is valid*

$$V[t+\eta, x+z, y_\xi(\cdot)] - V[t, x, y(\cdot)] = \frac{\partial V[p]}{\partial t}\eta + \frac{\partial V[p]}{\partial x}z +$$

$$+\,\partial_y V[p]\xi + o_Y\left(\sqrt{\|z\|^2 + \eta^2 + \xi^2}\right), \tag{11.10}$$

where $z \in \mathbb{R}^n$, $\eta \geq 0$, $\xi \in [0, \beta]$, *and, in general case* $o_Y(\cdot)$ *depends on the choice of* $Y(\cdot) \in E[x, y(\cdot)]$ $\left(\dfrac{o_Y(s)}{s} \to 0 \text{ as } s \to +0\right)$.

Note, in the definitions 11.5 and 11.6 properties of the invariant continuity and the invariant differentiability are defined simultaneously with respect to all variables $\{t, x, y(\cdot)\}$, and not separately by $y(\cdot)$. The invariant continuity and invariant differentiability of the functional (11.1) are defined by properties of the function $\Psi_Y(\eta, z, \xi)$ (11.2).

Hence by analogy with differentiability of finite-dimensional functions the following theorem is valid

Theorem 11.1 *In order for the functional (11.1) to be in-invariant differentiable at a point* $p \in \mathbb{R} \times \mathbb{R}^n \times Q[-\tau, 0)$ *it is necessary that the functional has partial derivatives* $\dfrac{\partial V[p]}{\partial t}$, $\dfrac{\partial V[p]}{\partial x}$, $\partial_y V[p]$, *and sufficient that these derivatives be invariantly continuous at the point* p.

Remark 11.1 One can easily verify that functionals, considered in previous sections, have invariantly continuous derivatives $\dfrac{\partial V[p]}{\partial t}$, $\dfrac{\partial V[p]}{\partial x}$ and $\partial_y V[p]$, and hence are invariantly differentiable.

11.4 Invariant derivative in the class $B[-\tau, 0]$

In example 11.3 the singular functional has the invariant derivative not at every function $y(\cdot) \in Q[-\tau, 0)$, but only for functions $y(\cdot)$ differentiable at the point $s = -\tau$. Consider two examples of calculating invariant derivatives of regular functionals.

Example 11.9 To calculate the invariant derivative of the functional

$$V[y(\cdot)] = \int_{-\tau}^{0} \beta(s, y(s))ds\,, \tag{11.11}$$

($\beta : [-\tau, 0] \times \mathbb{R}^n \to \mathbb{R}$) construct the function

$$\Psi_Y(\xi) = V[y_\xi(\cdot)] = \int_{-\tau}^{0} \beta(s, Y(\xi + s))ds \tag{11.12}$$

$\xi \in [0, \beta)$.

We have to consider three cases:

1) If the function $\beta(s, y)$ is only *continuous*, then the functional (11.11) does not have the invariant derivative at a functions $y(\cdot) \in Q[-\tau, 0)$, because in this case the function (11.12) does not have derivatives at the zero;

2) If the function $\beta(s, y)$ has continuous partial derivative $\frac{\partial \beta(s, y)}{\partial s}$, then the functional (11.11) has at every point $\{x, y(\cdot)\} \in H$ the invariant derivative

$$\partial_y V[x, y(\cdot)] = \left(\frac{\partial \Psi_Y}{\partial \xi} \right)_{\xi=+0} =$$

$$= \frac{d}{d\xi} \Big(\int_{-\tau}^{0} \beta(s, Y(\xi + s)) ds \Big)_{\xi=+0} =$$

$$= \frac{d}{d\xi} \Big(\int_{-\tau+\xi}^{\xi} \beta(s - \xi, Y(s)) ds \Big)_{\xi=+0} =$$

$$= \beta(0, Y(0)) - \beta(-\tau, Y(-\tau)) - \int_{-\tau}^{0} \frac{\partial \beta(s, Y(s))}{\partial s} ds =$$

$$= \beta(0, x) - \beta(-\tau, y(-\tau)) - \int_{-\tau}^{0} \frac{\partial \beta(s, y(s))}{\partial s} ds;$$

3) If the function $\beta(s, y)$ has the continuous partial derivative $\frac{\partial \beta(s, y)}{\partial y}$ with respect to y, and besides a function $y(\cdot) \in Q[-\tau, 0)$ and its continuation $Y(\cdot) \in E[x, y(\cdot)]$ have piece-wise continuous derivatives, then the functional (11.11) has at a point $\{x, y(\cdot)\} \in H$ the invariant derivative

$$\partial_y V[x, y(\cdot)] = \left(\frac{\partial \Psi_Y}{\partial \xi} \right)_{\xi=+0} =$$

$$= \frac{d}{d\xi}\Big(\int_{-\tau}^{0} \beta(s, Y(\xi+s))ds\Big)_{\xi=+0} =$$

$$= \int_{-\tau}^{0} \frac{\partial \beta'(s, y(s))}{\partial y} \dot{y}(s)ds\,. \qquad (11.13)$$

□

Example 11.10 Consider the functional defined by the Stiltjes integral

$$V[y(\cdot)] = \int_{-\tau}^{0} \beta(s, y(s))d\lambda(s)\,, \qquad (11.14)$$

where the function $\beta(s, y) : [-\tau, 0] \times \mathbb{R}^n \to \mathbb{R}$ has continuous partial derivative $\dfrac{\partial \beta(s, y)}{\partial y}$ with respect to the second variable y, and the function $\lambda : [-\tau, 0] \to \mathbb{R}$ is a function of bounded variation.

Remember that functional (11.14) is defined only on continuous functions $y(\cdot) \in C[-\tau, 0]$. To calculate the invariant derivative we have to construct the function

$$\Psi_Y(\xi) = V[y_\xi(\cdot)] = \int_{-\tau}^{0} \beta(s, Y(\xi+s))d\lambda(s) \qquad (11.15)$$

$\xi \in [0, \beta)$ and calculate its derivative at the aero. The function (11.15) has right derivative at the zero only if the function $y(\cdot)$ and its continuation $Y(\cdot) \in E[x, y(\cdot)]$ are continuous differentiable: in this case

$$\partial_y V[x, y(\cdot)] = \left(\frac{\partial \Psi_Y}{\partial \xi}\right)_{\xi=+0} =$$

$$= \frac{d}{d\xi}\Big(\int_{-\tau}^{0} \beta(s, Y(\xi+s))d\lambda(s)\Big)_{\xi=+0} =$$

$$= \int_{-\tau}^{0} \frac{\partial \beta'(s, y(s))}{\partial y} y(s) d\lambda(s) \,. \tag{11.16}$$

□

Note that functionals, defined by formulas (11.13) and (11.16), are not generally speaking invariant derivatives of the corresponding functionals (11.11) and (11.14) in the sense of the definition 11.1, because in the first case the continuation $Y(\cdot)$ should have piece-wise continuous derivative on $[0, \Delta)$, and in the second case the function $Y(\cdot)$ should be continuous differentiable on $[-\tau, \Delta)$, meanwhile in the definition 11.1 the function $Y(\cdot)$ is assumed be only continuous on $[0, \Delta)$.

Taking into account this fact, below we introduce a definition of the invariant derivative in a class of functions $B[-\tau, 0]$. For example such class $B[-\tau, 0]$ can be sets of sufficiently smooth functions $C^k[-\tau, 0]$.

For $h = \{x, y(\cdot)\} \in B[-\tau, 0]$ $\beta > 0$ denote:

$$\hat{E}_\beta[x, y(\cdot)] = \Big\{ Y(\cdot) \in E[h] : Y_\xi \in B[-\tau, 0], \xi \in [0, \beta] \Big\}$$

and $\hat{E}[h] = \bigcup\limits_{\beta > 0} \hat{E}_\beta[h]$.

Then formulas (11.13) and (11.16) define invariant derivatives of the corresponding functionals in the sense of the next definition

Definition 11.7 *Functional (11.1)*

1) *Has at point* $\{t, x, y(\cdot)\} \in \mathbb{R} \times B[-\tau, 0]$ *an invariant derivative* $\partial_y V$ *in class* $B[-\tau, 0]$, *if for any* $Y(\cdot) \in \hat{E}[x, y(\cdot)]$ *function (11.2) has at zero a finite right-hand side derivative* $\dfrac{d\Psi_Y(0)}{d\xi}$, *invariant with respect to* $Y(\cdot) \in \hat{E}[x, y(\cdot)]$;

2) Invariantly differentiable in class $B[-\tau, 0]$*, if for any* $\{t, x, y(\cdot)\} \in \mathbb{R} \times B[-\tau, 0]$ *and* $Y(\cdot) \in \hat{E}[x, y(\cdot)]$ *(11.10) is valid.*

The *invariant derivative in a class* $B[-\tau, 0]$ is also called a B–invariant derivative, for example, a C^1–invariant derivative if $B[-\tau, 0] = C^1[-\tau, 0]$.

Example 11.11 Let us show that the quadratic functional

$$V[y(\cdot)] = \int_{-\tau}^{0}\int_{-\tau}^{0} y'(s)\gamma[s, u]y(u)dsdu$$

is invariantly differentiable in class $H^1 = \mathbb{R}^n \times \times Q^1[-\tau, 0)$, if elements of $n \times n$ matrix $\gamma[s, u]$ are continuous on $[-\tau, 0] \times [-\tau, 0]$. Remember, $Q^1[-\tau, 0)$ – is the space of functions $y(\cdot) \in Q[-\tau, 0)$, which have piece-wise continuous derivatives. Fix a point $p = \{t, x, y(\cdot)\} \in \mathbb{R} \times H^1$ and for a continuation $Y(\cdot) \in \hat{E}[x, y(\cdot)]$ consider the function

$$\Psi_Y(\xi) = \int_{-\tau}^{0}\int_{-\tau}^{0} Y'(s+\xi)\gamma[s, u]Y(u+\xi)dsdu\,.$$

Taking into account differentiability of $Y(\cdot)$ one can calculate

$$\partial_y V[x, y(\cdot)] = \left(\frac{\partial \Psi_Y}{\partial \xi}\right)_{\xi=+0} =$$

$$= \int_{-\tau}^{0}\int_{-\tau}^{0} \dot{Y}'(s)\gamma[s, u]Y(u)dsdu + \int_{-\tau}^{0}\int_{-\tau}^{0} Y'(s)\gamma[s, u]\dot{Y}(u)dsdu$$

$$= \int_{-\tau}^{0}\int_{-\tau}^{0} \dot{y}'(s)\,\gamma[s, u]y(u)dsdu + \int_{-\tau}^{0}\int_{-\tau}^{0} y'(s)\gamma[s, u]\dot{y}(u)dsdu\,.$$

Note, if elements of the matrix $\gamma[s,u]$ are continuous differentiable, then the functional $V[y(\cdot)]$ is invariantly differentiable on $\mathbb{R}^n \times Q[-\tau,0)$, because previous formula can be continued by the following way:

$$\partial_y V[x,y(\cdot)] = \left(\frac{\partial \Psi_Y}{\partial \xi}\right)_{\xi=+0} =$$

$$= \int_{-\tau}^{0} x'\,\gamma[0,u]y(u)du - \int_{-\tau}^{0} y'(-\tau)\gamma[-\tau,u]y(u)du -$$

$$- \int_{-\tau}^{0}\int_{-\tau}^{0} y'(s)\frac{\partial \gamma[s,u]}{\partial s}y(u)dsdu + \int_{-\tau}^{0} y'(s)\gamma[s,0]xds -$$

$$- \int_{-\tau}^{0} y'(s)\gamma[s,-\tau]y(-\tau)ds - \int_{-\tau}^{0}\int_{-\tau}^{0} y'(s)\frac{\partial \gamma[s,u]}{\partial u}y(u)dsdu\,.$$

□

Example 11.12 The singular functional (9.4) has the *invariant derivative*

$$\partial_y V[\dot{y}(\cdot)] = \frac{\partial P'[y(-\tau)]}{\partial y}y(-\tau) \qquad (11.17)$$

in class $\mathbb{R}^n \times Q^1[-\tau,0)$, if the function $P[\cdot] : \mathbb{R}^n \to \mathbb{R}$ is continuous differentiable. To prove this fact fix a pair $\{x, y(\cdot)\} \in \mathbb{R}^n \times Q^1[-\tau,0)$ and $\beta \in (0,\tau)$. Because the function $\Psi_Y(\xi) \equiv P[Y(\xi-\tau)] = P[y(\xi-\tau)]$ does not depend on values of the function $Y(\cdot) \in \hat{E}_\beta[x,y(\cdot)]$ on $[0,\beta]$, hence the derivative

$$\left.\frac{d\Psi_Y(\xi)}{d\xi}\right|_{\xi=+0} = \left.\frac{dP'[y(\xi-\tau)]}{d\xi}\right|_{\xi=+0} =$$

$$= \frac{\partial P'[y(-\tau)]}{\partial y}\left.\frac{dy(\xi-\tau)}{d\xi}\right|_{\xi=+0} = \frac{\partial P'[y(-\tau)]}{\partial y}y(-\tau)$$

also does not depend on the values of $Y(\cdot)$ on $[0,\beta]$. Thus, functional (11.17) is the invariant derivative of the functional (9.4) in the class $\mathbb{R}^n \times Q^1[-\tau, 0)$.

Note that the invariant derivative (11.17) depends on the value of the derivative of the function $y(\cdot)$ only at the point $s=-\tau$, hence generally speaking this functional is the invariant derivative of the functional (9.4) in the class of functions $y(\cdot)$ which have right-hand derivatives only at poin $s=-\tau$. □

12 Coinvariant derivative

12.1 Coinvariant derivative of functionals

In specific problems (for example, calculating a functional $V[t,x,y(\cdot)]$ along solutions) variables η and ξ have a similar (common) "physical" sense – to characterize shift with respect to time. For this reason, in many cases it is convenient consider an increment with respect to t and an increment with respect to $Y(\cdot)$ adjusted (i.e. $\eta=\xi$). Then the function (11.2) has the form

$$\hat{\Psi}_Y(\eta,z)=V[t+\eta,x+z,y_\eta(\cdot)]\,, \tag{12.1}$$

$\eta \in [0,\beta]$, $z\in \mathbb{R}^n$.

Definition 12.1 *A functional $V[t,x,y(\cdot)]$ has at a point $p=\{t,x,y(\cdot)\}\in \mathbb{R}\times\mathbb{R}^n\times Q[-\tau,0)$ the coinvariant derivative $\partial_t V$, if for every $Y(\cdot)\in E[x,y(\cdot)]$ the function (12.1) has at zero the right-hand derivative by η, invariant with respect to $Y(\cdot)$. In this case*

$$\partial_t V=\left.\frac{\partial\hat{\Psi}_Y(\eta,z)}{\partial\eta}\right|_{\eta=+0,z=0}.$$

One can easily prove the following proposition

Theorem 12.1 *If the functional (11.1) has at a point* $p = \{t, x, y(\cdot)\}$ *the partial derivative* $\frac{\partial V[p]}{\partial t}$ *and the invariant derivative* $\partial_y V[p]$*, then the functional has at the point* p *the coinvariant derivative* $\partial_t V[p]$*, moreover,*

$$\partial_t V[p] = \frac{\partial V[p]}{\partial t} + \partial_y V[p] .$$

Example 12.1 It follows from the theorem 12.1 that the functional (11.4) has the coinvariant derivative

$$\partial_t V[t, x, y(\cdot)] = \frac{\partial g(t, x)}{\partial t} + \alpha(x) - \alpha(y(-\tau)) .$$

□

It is shown in [18] that a proposition inverse to the theorem 12.1 is not valid.

The following example shows that a proposition converse to the theorem 12.1 is not generally speaking valid.

Example 12.2 Consider the functional

$$V[t, y(\cdot)] = \int_{-t}^{0} \int_{t}^{\infty} k(\nu, t + s) y^2(s) d\nu ds , \tag{12.2}$$

where $k : \mathbb{R} \times \mathbb{R} \to \mathbb{R}$ is a continuous function such that

$$\int_{-t}^{0} \int_{t}^{\infty} k(\nu, \eta) d\nu d\eta < +\infty$$

and a function

$$\phi(t, \eta) = \int_{t}^{\infty} k(\nu, \eta) d\nu$$

is defined and continuous for $0 \le \eta \le t < +\infty$.

The functional (12.2) is defined on $\mathbb{R} \times Q(-\infty, 0)$, however it does not have the partial derivative with respect to t nor the invariant derivative by $y(\cdot)$, because the continuity of $k(\cdot,\cdot)$ it is not sufficient for existence of these derivatives. However one can easily verify that the functional (12.2) has at the point $\{t, x, y(\cdot)\} \in \mathbb{R} \times \mathbb{R}^n \times Q(-\infty, 0)$ the coinvariant derivative

$$\partial_t V[t, x, y(\cdot)] = x^2 \int_t^{\infty} k(\nu, t) d\nu - \int_{-t}^{0} k(t, t+s) y^2(s) ds .$$

□

Similar to definitions 11.5 11.6 one can introduce notions of *coinvariant continuity* (*ci*–continuity) and *coinvariant differentiability* (*ci*–differentiability) of functionals

Definition 12.2 *The Functional (11.1) is at a point* $p = \{t, x, y(\cdot)\} \in \mathbb{R} \times \mathbb{R}^n \times Q[-\tau, 0)$

1) *Coinvariantly continuous (ci–continuous), if for every* $Y(\cdot) \in E[x, y(\cdot)]$ *the corresponding function (12.1) is continuous at zero (continuity with respect to* η *from the right) and the limit value* $\hat{\Psi}_Y(0)$ *is invariant with respect* $Y(\cdot)$;

2) *Coinvariantly differentiable (ci–differentiable), if at the point exist finite derivatives* $\partial_t V[p]$ *and* $\dfrac{\partial V[p]}{\partial x}$, *and for every function* $Y(\cdot) \in E[x, y(\cdot)]$ *the following equality is valid*

$$V[t+\eta, x+z, y_\eta(\cdot)] - V[t, x, y(\cdot)] =$$
$$= \partial_t V[p]\eta + \frac{\partial V[p]}{\partial x} z + o_Y\left(\sqrt{\|z\|^2 + \eta^2}\right)$$

for $z \in \mathbb{R}^n$, $\eta \in [0, \beta]$; *function* $o_Y(\cdot)$ *depends on the choice of* $Y(\cdot) \in E[x, y(\cdot)]$ $\left(\dfrac{o_Y(\sqrt{s})}{\sqrt{s}} \to 0 \text{ as } s \to +0\right)$.

Further we will also utilized the following notions of continuity

Definition 12.3 *Functional (11.1) is:*

1) *y–continuous on shift at a point* $h^*\{t_*, x^*, y^*(\cdot)\} \in \mathbb{R} \times \mathbb{R}^n \times Q[-\tau, 0)$, *if* $(\exists \beta > 0)$ $(\forall Y(\cdot) \in E_\beta[h^*])$ *the function* $\phi_Y(\eta) = V[t_*, x^*, Y_\eta]$ *is continuous on* $[0, \beta]$;

2) *t–continuous on shift at a point* $\{t_*, x^*, y^*(\cdot)\} \in \mathbb{R} \times \mathbb{R}^n \times Q[-\tau, 0)$, *if* $(\exists \beta > 0)$ $(\forall Y(\cdot) \in E_\beta[h^*])$ *the function* $\Phi_Y(t) = V[t, x^*, Y_{t-t_*}]$ *is continuous on* $[t_*, t_*+\beta]$;

3) *Continuous on shift at a point* $\{t_*, x^*, y^*(\cdot)\} \in \mathbb{R} \times \mathbb{R}^n \times Q[-\tau, 0)$, *if* $(\exists \beta > 0)$ $(\forall Y(\cdot) \in E_\beta[h^*])$ *the function* $\Psi_Y(t) = V[t, Y(t), Y_{t-t_*}]$ *is continuous on* $[t_*, t_* + \beta]$.

12.2 Coinvariant derivative in a class $B[-\tau, 0]$

In this section we introduce the notion of the coinvariant derivative of a functional

$$V[t, x, y(\cdot)] : \mathbb{R} \times \mathbb{R}^n \times B[-\tau, 0) \to \mathbb{R} \tag{12.3}$$

at points $p = \{t, h\} \in \mathbb{R} \times B[-\tau, 0]$.

Remember that $B[-\tau, 0]$ $(B[-\tau, 0))$ denotes a space of sufficiently smooth functions of $C[-\tau, 0]$ $(C[-\tau, 0))$, and the inclusion $h \equiv \{x, y(\cdot)\} \in B[-\tau, 0]$ means that $x = \lim\limits_{s \to -0} y(s)$.

Note, further we consider coinvariant differentiability of the functional (12.3) on $\mathbb{R} \times B[-\tau, 0]$, though the functional is defined on more wide sets. This allows us to simplify some constructions [15].

Definition 12.4 *The functional (12.3)*

---

[15]In general case one can define coinvariant derivative of functional on $\mathbb{R} \times \mathbb{R}^n \times B[-\tau, 0)$

1) *Has at a point* $\{t, h\} \in \mathbb{R} \times B[-\tau, 0]$ *the coinvariant derivative* $\partial_t V$ *in the class* $B[-\tau, 0]$, *if for every* $Y(\cdot) \in \hat{E}[h]$ *the corresponding function (12.1) has at the zero the right-hand derivative* $\frac{d\hat{\Psi}_Y(0,0)}{d\eta}$ *Which is invariant with respect to* $Y(\cdot) \in \hat{E}[h]$;

2) Invariantly differentiable *in the class* $B[-\tau, 0]$, *if for every* $\{t, h\} \in \mathbb{R} \times B[-\tau, 0]$ *and* $Y(\cdot) \in \hat{E}[h]$ *the formula (11.10) is valid.*

Coinvariant derivative in class $B[-\tau, 0]$, will be also called B–coinvariant derivative, for example C^1–coinvariant derivative, if $B[-\tau, 0] = C^1[-\tau, 0]$.

Definition 12.5 *The functional (12.3) is at a point* $p = \{t, h\} \in \mathbb{R} \times B[-\tau, 0]$

1) coinvariantly continuous in a class $B[-\tau, 0]$, *if for every* $Y(\cdot) \in \hat{E}[h]$ *the corresponding function (12.1) is continuous at the zero (continuity by* η *from the right) and the limit value* $\hat{\Psi}_Y(0)$ *is invariant with respect to* $Y(\cdot) \in \hat{E}[h]$;

2) coinvariantly differentiable in class $B[-\tau, 0]$, *if there exist finite coinvariant derivative* $\partial_t V[p]$ *in class* $B[-\tau, 0]$ *and partial derivative*
$\frac{\partial V[p]}{\partial x}$, *Such that for every* $Y(\cdot) \in \hat{E}[h]$

$$V[t + \eta, x + z, y_\eta(\cdot)] - V[t, x, y(\cdot)] =$$

$$= \partial_t V[p]\eta + \frac{\partial V[p]}{\partial x} z + o_Y\left(\sqrt{\|z\|^2 + \eta^2}\right)$$

for $z \in \mathbb{R}^n$, $\eta \in [0, \beta]$, *the function* $o_Y(\cdot)$ *can depend on choice of* $Y(\cdot) \in E[x, y(\cdot)]$ $\left(\frac{o_Y(\sqrt{s})}{\sqrt{s}} \to 0 \text{ as } s \to +0\right)$.

Example 12.3 To calculate the invariant and coinvariant derivatives of the singular functional (10.9) at a point $\{x, y(\cdot)\} \in \mathbb{R}^n \times Q^1[-\tau, 0)$ take $\beta \in (0, \tau)$ and consider the function (11.2)

$$\Psi_{_Y}(\eta, \xi) \equiv P[Y(\xi - \tau^*(t + \eta))] = P[y(\xi - \tau^*(t + \eta))],$$

which does not depend on values of $Y(\cdot) \in \hat{E}_\beta[x, y(\cdot)]$ on $[0, \beta]$.

If $P : \mathbb{R}^n \to \mathbb{R}$ is a continuous differentiable function, then one can calculate the invariant derivative

$$\partial_y V[\dot{y}(\cdot)] = \left.\frac{\partial \Psi_{_Y}(0, \xi)}{\partial \xi}\right|_{\xi=+0} = \left.\frac{dP[y(\xi - \tau^*(t))]}{d\xi}\right|_{\xi=+0} =$$

$$= \left.\frac{\partial P'[y(-\tau^*(t))]}{\partial y} \frac{dy(\xi - \tau^*(t))}{d\xi}\right|_{\xi=+0} =$$

$$= \frac{\partial P'[y(-\tau^*(t))]}{\partial y} y(-\tau^*(t))$$

and the coinvariant derivative

$$\partial_t V[\dot{y}(\cdot)] = \left.\frac{d\Psi_{_Y}(\xi, \xi)}{d\xi}\right|_{\xi=+0} = \left.\frac{dP[y(\xi - \tau^*(t + \xi))]}{d\xi}\right|_{\xi=+0} =$$

$$= \left.\frac{\partial P'[y(-\tau^*(t))]}{\partial y} \frac{dy(\xi - \tau^*(t + \xi))}{d\xi}\right|_{\xi=+0} =$$

$$= \frac{\partial P'[y(-\tau^*(t))]}{\partial y} \cdot$$

$$\cdot \left(\dot{y}(-\tau^*(t)) - \dot{y}(-\tau^*(t)) \left.\frac{d\tau^*(t + \xi)}{d\xi}\right|_{\xi=+0} \right) =$$

$$= \frac{\partial P'[y(-\tau^*(t))]}{\partial y} \dot{y}(-\tau^*(t)) \left[1 - \dot{\tau}^*(t)\right]$$

in the class $\mathbb{R}^n \times Q^1[-\tau, 0)$. □

12.3 Properties of the coinvariant derivative

For coinvariantly differentiable functionals the basic rules of the finite dimensional calculus are valid.

Theorem 12.2 *If functional $V[t,h]$, $W[t,h] : \mathbb{R}\times B[-\tau,0] \to \mathbb{R}$ have at a point $p = \{t,h\} \in \mathbb{R}\times B[-\tau,0]$ coinvariant derivatives $\partial_t V[t,h]$, $\partial_t W[t,h]$ in class $B[-\tau,0]$, then the sum and the multiplication of these functionals have at the point p coinvariant derivatives in the class $B[-\tau,0]$ and*

$$\partial_t\Big(V[t,h] + W[t,h]\Big) = \partial_t V[t,h] + \partial_t W[t,h], \qquad (12.4)$$

$$\partial_t\Big(V[t,h]\ W[t,h]\Big) = \partial_t V[t,h]\ W[t,h] + V[t,h]\ \partial_t W[t,h]. \qquad (12.5)$$

If additionally $W[t,h] \neq 0$ then

$$\partial_t\left(\frac{V[t,h]}{W[t,h]}\right) = \frac{\partial_t V[t,h]\, W[t,h] - V[t,h]\partial_t W[t,h]}{W^2[t,h]}. \qquad (12.6)$$

Proof. Fix a function $Y(\cdot) \in \hat{E}_\kappa[h]$ and construct

$$\Psi_1(\eta) \equiv W[t+\eta, Y_\eta], \quad \Psi_2(\eta) \equiv W[t+\eta, Y_\eta],$$

$\eta \in [0,\kappa]$. The sum of these functions has at the point $\eta = 0$ the partial derivative

$$\frac{d}{d\eta}\Big(\Psi_1(\eta) + \Psi_2(\eta)\Big)\Big|_{\eta=0} = \dot{\Psi}_1(0) + \dot{\Psi}_2(0) =$$

$$= \partial_t f(t,h) + \partial_t g(t,h)$$

which do not depend on a specific form of the function $Y(\cdot)$. Hence the sum of the functionals V and W is invariantly differentiable in class $B[-\tau,0]$ functional and the formula (12.4) is valid.

Similar one can prove formulas (12.5) and (12.6). □

Remark 12.1 Obviously, the coinvariant derivative of a *constant* functional is equal to the zero.

Remark 12.2 It follows from the formular (12.5) that $\partial_t\Big(c \cdot V[t,h]\Big) = c \cdot \partial_t V[t,h]$ if the functional $V[t,h] : \mathbb{R} \times B[-\tau,0] \to \mathbb{R}$ is coinvariantly differentiable at the point $\{t,h\} \in \mathbb{R} \times B[-\tau,0]$ and $c = const$.

Coinvariant derivative of the composition of a *function* and a *functional* can be calculated by the following rule.

Theorem 12.3 *If a functional $V[t,h] : \mathbb{R} \times B[-\tau,0] \to \mathbb{R}$ is coinvariantly differentiable in a class $B[-\tau,0]$ at a point $p^* = \{t,h\} \in \mathbb{R} \times B[-\tau,0]$ and a function $w(\zeta) : \mathbb{R} \to \mathbb{R}$ is differentiable at the point $\zeta^* = V[p^*]$, then the functional $g(t,h) = w(V[t,h])$ is coinvariantly differentiable in the class $B[-\tau,0]$ at point p^* and*

$$\partial_t g(p^*) = \frac{d\,w(V[p^*])}{d\zeta}\partial_t V[p^*]\,. \tag{12.7}$$

For example, if a functional $V[t,h] : \mathbb{R} \times B[-\tau,0] \to \mathbb{R}$ is coinvariantly differentiable at a point $\{t,h\} \in \mathbb{R} \times B[-\tau,0]$, then

$$\partial_t\Big(c \cdot V[t,h]\Big) = c \cdot \partial_t V[t,h]\,,$$

$$\partial_t\Big(V^k(t,h)\Big) = kV^{k-1}(t,h) \cdot \partial_t V[t,h] \quad (k - const)\,,$$

$$\partial_t\Big(a^{V[t,h]}\Big) = a^{V[t,h]} \cdot \ln a \cdot \partial_t V[t,h]\,,$$

$$\partial_t\Big(e^{V[t,h]}\Big) = e^{V[t,h]}\,\partial_t V[t,h]\,,$$

$$\partial_t\Big(\log_a V[t,h]\Big) = \frac{\partial_t V[t,h]}{V[t,h]\,\ln a}\,,$$

$$\partial_t\Big(\ln V[t,h]\Big) = \frac{\partial_t V[t,h]}{V[t,h]},$$

$$\partial_t\Big(\sin V[t,h]\Big) = \cos V[t,h]\cdot\partial_t V[t,h],$$

$$\partial_t\Big(\cos V[t,h]\Big) = -\sin V[t,h]\cdot\partial_t V[t,h],$$

$$\partial_t\Big(\arcsin V[t,h]\Big) = \frac{\partial_t V[t,h]}{\sqrt{1-V^2[t,h]}},$$

$$\partial\Big(\arctan V[t,h]\Big) = \frac{\partial_t V[t,h]}{1+V^2[t,h]}.$$

12.4 Partial derivatives of high order

Calculate for a functional the partial derivatives of a high order, where

$$Q^m[-\tau,0) \equiv \left\{ y(\cdot)\in Q[-\tau,0) \;\middle|\; \frac{d^k y(s)}{ds^k}\in Q[-\tau,0) \text{ for } k=1,\ldots,m \right\}.$$

In this section $B[-\tau,0) = Q^m[-\tau,0)$.

Let us assume that a functional

$$V[t,x,y(\cdot)] : \mathbb{R}\times\mathbb{R}\times Q^m[-\tau,0)\to\mathbb{R} \tag{12.8}$$

is Q^m–coinvariantly differentiable at every point $p = \{t,x,y(\cdot)\}$. Then its derivatives $\partial_t V[t,x,y(\cdot)]$ and $\dfrac{\partial V[t,x,y(\cdot)]}{\partial x}$ are functional and vector-functional on $\mathbb{R}\times\mathbb{R}^n\times Q^{m+1}[-\tau,0)$ respectively, i.e. in general case for calculating invariant derivatives of high order smoothness of functions $y(\cdot)$ should increase.

If $\partial_t V$ and $\dfrac{\partial V}{\partial x}$ have partial derivatives then these (second) partial derivatives are denoted by

$$\frac{\partial}{\partial x}\left(\frac{\partial V[t,x,y(\cdot)]}{\partial x}\right) = \frac{\partial^2 V[t,x,y(\cdot)]}{\partial x^2},$$

$$\partial_t\Big(\partial_t V[t,x,y(\cdot)]\Big) = \partial_t^2 V[t,x,y(\cdot)]\,,$$

$$\frac{\partial}{\partial x}\Big(\partial_t V[t,x,y(\cdot)]\Big)\,,\quad \partial_t\left(\frac{\partial V[t,x,y(\cdot)]}{\partial x}\right).$$

In a similar way one can define high order derivatives. For example

$$\partial_t^2\left(\frac{\partial V[t,x,y(\cdot)]}{\partial x}\right)$$

is the third order derivative of the functional V, calculated one time with respect to x and twice as the coinvariant derivative.

Emphasized once more, we assume that the coinvariant differentiation increases at one time the requirement on smoothness of the basic space functions $B[-\tau,0)$, i.e. if functional V is defined on $Q^m[-\tau,0)$, then it is assumed that its coinvariant derivative is defined on $Q^{m+1}[-\tau,0)$. Due to the structure of specific functionals, this assumption is valid in most of cases.

Example 12.4 Consider the functional

$$V[y(\cdot)] = \int_{-\tau}^{0} \alpha(y(s))ds\,, \tag{12.9}$$

where $\alpha(\cdot)$ is a continuous differentiable function. The coinvariant derivative of the functional is

$$\partial_t V[x,y(\cdot)] = \alpha(x) - \alpha(y(-\tau))\,.$$

The second coinvariant derivative

$$\partial_t^2 V[y(\cdot)] = \dot{\alpha}(y(-\tau))\,\dot{y}(-\tau)$$

is already defined on $Q^1[-\tau,0)$, meanwhile the third coinvariant derivative

$$\partial_t^3 V[y(\cdot)] = \ddot{\alpha}(y(-\tau))\;(\dot{y}(-\tau))^2 + \dot{\alpha}(y(-\tau))\,\ddot{y}(-\tau)$$

is defined on $Q^2[-\tau,0)$. □

Note, in general cases it is important to have an order of calculating partial derivatives.

Example 12.5 If the functional

$$V[y(\cdot)] = \int_{-\tau}^{0} y(s)ds \tag{12.10}$$

has partial derivatives

$$\frac{\partial V[y(\cdot)]}{\partial x} = 0 \,, \ \partial_t V[x, y(\cdot)] = x - y(-\tau) \,,$$

then

$$\partial_t \left(\frac{\partial V[y(\cdot)]}{\partial x} \right) = 0 \quad \text{and} \quad \frac{\partial}{\partial x} \Big(\partial_t V[x, y(\cdot)] \Big) = 1 \,.$$

Therefore

$$\partial_t \left(\frac{\partial V}{\partial x} \right) = 0 \neq \frac{\partial}{\partial x} \Big(\partial_t V \Big) \,.$$

□

12.5 Formulas of i–smooth calculus for mappings

In this section we consider formulas of i-smooth calculus for *mappings*

$$f(t, x, y(\cdot)) : \mathbb{R} \times \mathbb{R}^n \times Q[-\tau, 0) \to \mathbb{R}^n \tag{12.11}$$

or in the coordinate form

$$f(t, x, y(\cdot)) = \Big(f_1(t, x, y(\cdot)), \ldots, f_n(t, x, y(\cdot)) \Big)'$$

composed of functionals $f_i(t, x, y(\cdot)) : \mathbb{R} \times \mathbb{R}^n \times Q[-\tau, 0) \to \mathbb{R}$, $i = 1, \ldots, n$. The mapping (12.11) is also called *vector-functional.*

The *partial derivative with respect to* t, the *invariant derivative* and the *coinvariant derivative* of the mapping (12.11) are the following mappings

$$\frac{\partial f(t,x,y(\cdot))}{\partial t} = \left(\frac{\partial f_1(t,x,y(\cdot))}{\partial t}, \ldots, \frac{\partial f_n(t,x,y(\cdot))}{\partial t}\right)',$$

$$\partial_y f(t,x,y(\cdot)) = \Big(\partial_y f_1(t,x,y(\cdot)), \ldots, \partial_y f_n(t,x,y(\cdot))\Big)'$$

and

$$\partial_t f(t,x,y(\cdot)) = \Big(\partial_t f_1(t,x,y(\cdot)), \ldots, \partial_t f_n(t,x,y(\cdot))\Big)',$$

respectively.

The *partial derivative* of the mapping (12.11) *with respect to the vector* $x = (x_1, \ldots, x_n)'$ is the matrix

$$\frac{\partial f(t,x,y(\cdot))}{\partial x} = \left(\frac{\partial f_i(t,x,y(\cdot))}{\partial x_j}\right)_{n\times n}.$$

13 Brief overview of Functional Differential Equation theory

The section is not a regular introduction to FDE theory. We assume that the reader is already familiar with the fundamentals of time-delay system theory and that this section explains for him/her some of the notions that is used in the book.

13.1 Functional Differential Equations

Functional Differential Equation

$$\dot{x}(t) = f(t, x(t), x_t(\cdot)) \tag{13.1}$$

($f : \mathbb{R} \times \mathbb{R}^n \times Q[-\tau, 0) \to \mathbb{R}^n$) is a generalization of the ordinary differential equation

$$\dot{x}(t) = g(t, x(t)) \tag{13.2}$$

($g : \mathbb{R} \times \mathbb{R}^n \to \mathbb{R}^n$) for the case, when at a moment t the velocity $\dot{x}(t)$ of the process $x(t)$ besides the current state $x(t)$, depends also on pre-history $x_t(\cdot) = \{x(t+s), -\tau \leq s < 0\}$.

For example, the equation

$$\dot{x}(t) = G[t, x(t), x(t - \tau(t))] \qquad (13.3)$$

($G : \mathbb{R} \times \mathbb{R}^n \times \mathbb{R}^n \to \mathbb{R}^n\,,\;\; 0 \leq -\tau(t) \leq -\tau$)

describes the system (process) with *discrete* delay, and the equation

$$\dot{x}(t) = \int\limits_{-\tau}^{0} \Phi[t, s, x(t+s)]ds \qquad (13.4)$$

($\Phi : \mathbb{R} \times [-\tau, 0] \times \mathbb{R}^n \to \mathbb{R}^n$) is the system with *distributed* delay.

The structure of specific functional differential equations is defined by the structure of functional $f(\cdot, \cdot, \cdot)$ in the right-hand side of the system (13.1).

Moreover, as a rule, in the structure of these functionals one can separate the finite dimensional components t and x, and the functional component $y(\cdot)$. For this reason one can separate in the structure of FDE the finite-dimensional vector $x(t)$ (characterizing *finite-dimensional* properties of the system) and the function-delay $x(t+s)$, $-\tau \leq s < 0$ (characterizing *functional(infinite-dimensional)* properties of the system.

For mathematical describing functional differential equations the following representations are usually used

$$\dot{x}(t) = f(t, x(t), x(t+s))\,, \;\; -\tau \leq s < 0\,, \qquad (13.5)$$

and

$$\dot{x}(t) = F[t, x(t+\hat{s})]\,, \;\; -\tau \leq \hat{s} \leq 0\,, \qquad (13.6)$$

($F : \mathbb{R} \times Q[-\tau, 0] \to \mathbb{R}^n$).

Note that representations (13.1), (13.5), (13.6) are equivalent, and $f : \mathbb{R} \times \mathbb{R}^n \times Q[-\tau, 0) \to \mathbb{R}^n$ is the operator $F : \mathbb{R} \times Q[-\tau, 0] \to \mathbb{R}^n$, acting in the space $\mathbb{R} \times H$. [16]

Sometimes we will use the term function for a pair $\{x, y(\cdot)\} \in H$. It is convenient to use the representation of the space $Q[-\tau, 0]$ as $H = \mathbb{R}^n \times Q[-\tau, 0)$, because for FDE the point $q(0)$ and points $\{q(s)$ play generally speaking different roles. So we use different symbols for the point $q(0)$ and other points $\{q(s), \tau \leq s < 0\}$ of the function $q(\cdot) \in Q[-\tau, 0] : x = q(0)$ and $y(\cdot) = \{q(s), \tau \leq s < 0\}$. Moreover, as a rule, we consider x and $y(\cdot)$ as independent components of the FDE state (remember that $H = \mathbb{R}^n \times Q[-\tau, 0)$ and $Q[-\tau, 0]$ are isometric spaces. More accurately: the mapping f is the superposition of the mappings F and π^{-1}, i.e. $f(t, h) = F[t, \pi^{-1}(h)]$, $(t, h) \in \mathbb{R} \times H$ (here π^{-1} – inverse mapping to the isometry $\pi : Q[-\tau, 0] \to H$).

13.2 FDE types

The section contains several simple examples of functional differential equations, for which solutions can be found in explicit form. These equations will be used for comparison of the exact solutions with approximate solutions obtained by numerical methods.

Example 13.1 Consider the equation

$$\dot{x}(t) = x(t - \tau), \tag{13.7}$$

($t \geq 0$, $\tau = 1$, x is the scalar variable) with the initial condition $x(0) \equiv 1$. To describe the unique solution of equation (13.7) for $t > 0$ insufficient knowing only initial

[16] Spaces H and $Q[-\tau, 0]$ are *isometric*: the corresponding isometry mapping $\pi : Q[-\tau, 0] \to H$ transfers functions $q(\cdot) \in Q[-\tau, 0]$ into pairs $\{q(0); q(s), \tau \leq s < 0\} \in H$.

point $x(0)$, it is necessary to define an initial function (prehistory) $y^0(\cdot)$. Let, for example, $y^0(s) = 1$ for $-1 \leq s < 0$. Then:

a) For $0 \leq t \leq 1$ we know the function $x(t-1) = y^0(t-1) = 1$, and, integrating the original equation we obtain

$$x(t) = x(0) + \int_0^t x(t-1)dt = 1 + t\,,$$

i.e. on this interval the solution is the linear function.

b) For $1 \leq t \leq 2$ we have $x(t-1) = 1 + (t-1) = t$, and integrating the original equation we obtain

$$x(t) = x(1) + \int_1^t x(t-1)dt = 2 + \int_1^t tdt = \frac{t^2}{2} + \frac{3}{2}\,,$$

i.e. on this interval the solution is the parabola.

Similar to the above described steps one can derive the solution on the next intervals of the unit length. In this case on the next intervals the smoothness of the solution increases. The applied method of integrating equations with delays is called *the step method.*

Note, in general, the solution of the more general equation

$$\dot{x}(t) = a\,x(t-\tau) \tag{13.8}$$

$(a > 0)$ with a constant initial function

$$y^0(s) = c\,,\ -\tau \leq s \leq 0 \tag{13.9}$$

$(c \in \mathbb{R})$ has on the interval $[0, \infty)$ the form

$$x(t) = c \sum_{m=0}^{[\frac{t}{\tau}]+1} \frac{a^m \Big(t - (m-1)\,\tau\Big)^m}{m!}\,. \tag{13.10}$$

Here $[\zeta]$ is the syllable (the integer part) of the number $\zeta \in \mathbb{R}$. □

Example 13.2 Consider the equation

$$\dot{x}(t) = x(t - (e^{-t} + 1)) + \cos t - \sin(t - e^{-t} - 1),$$

where $t \geq 0$, x is the scalar variable. The varying delay is $\tau(t) = e^{-t} + 1$, hence $1 < \tau(t) \leq 2$.

Take an initial function $y^0(t) = \sin t$ for $-2 \leq t \leq 0$. Then by the direct substitution it is verified that the function $x(t) = \sin t$ for $t \geq 0$ is the solution, corresponding to the initial function. □

Example 13.3 Consider the equation

$$\dot{x}(t) = \int_{-1}^{0} x_t(s)ds = \int_{t-\tau}^{t} x(\xi)d\xi \,. \tag{13.11}$$

Differentiating both sides of the equation, we obtain

$$\ddot{x}(t) = x(t) - x(t-\tau) \,. \tag{13.12}$$

This is the linear equation, so assuming that its solution has the form $x = e^{\lambda t}$, and substituting it into the equation we obtain for λ the transcendental characteristic equation

$$\lambda^2 - 1 = -e^{-\lambda\tau} \,,$$

which has a nonzero real root λ_0. This root can be found numerically: for example for $\tau = 1$: $\lambda_0 \approx 0.7$.

Therefore, if the initial condition is given in the form $x(s) = e^{\lambda_0 s}$, $s \in [-1, 0]$, then for $t \geq 0$ the solution of the equation (13.11) be the function $x(t) = e^{\lambda_0 t}$. □

We consider that the above examples 13.1 - 13.3 are good tests for testing numerical methods, because in these examples exact solutions are derived in an explicit form.

However for most of the functional differential equations finding exact solutions in their explicit form is an almost unsolvable problem, so the elaboration of effective numerical methods is one of the principle problem FDE theory.

13.3 Modeling by FDE

Functional differential equations describe various phenomena of mature and technical systems. Here we mention just a few most close to author's applications.

A typical object of the *automatic control theory* is the linear control system

$$\dot{x}(t) = A(t)x(t) + B(t)u(t) .$$

Here x is the phase vector and u is the control vector. The simplest and most effective is the linear control $u(t) = R(t)x(t)$ with a specially designed matrix $R(t)$. For almost two centuries these types of controls have been applied in real systems. Note that the transmission of information about the state $x(t)$ case a delay[17] τ, so in practice the control law has the form $u(t) = R(t)x(t-\tau)$ and there arises the delay in the original system, which becomes the system with the discrete delay

$$\dot{x}(t) = A(t)x(t) + B(t)R(t)x(t-\tau) .$$

The application of the proportional-integral regulator

$$u(t) = R(t)x(t) + \int_{-\tau}^{0} Q(t,s)x(t+s)ds ,$$

leads to the system with distributed delay

$$\dot{x}(t) = A(t)x(t) + BR(t)x(t) + B\int_{-\tau}^{0} Q(t,s)x(t+s)ds .$$

13.4 Phase space and FDE conditional representation

In distinction on ODE, to define a solution of FDE (13.1) it is necessary at every moment t to know the vector $x(t)$ and the function pre-history

$$x_t(\cdot) = \{x(t+s),\ -\tau \le s < 0\} .$$

So the phase space of FDE (13.1) should not be the finite dimensional space $\mathbb{R}^n$, but a *functional* space.

[17]a time necessary for measuring and transmitting the information

In this book we consider $H = \mathbb{R}^n \times Q[-\tau, 0)$ as the main phase space of the FDEs . This space is convenient for separating the finite dimensional and infinite dimensional components in FDE structure.

Note, the linear FDE are often considered in the Hilbert space $\mathbb{R}^n \times L_2[-\tau, 0)$.

Remark 13.1 The space H is *incomplete* space, however, as a rule, this fact does not cause problems, because after passing the time equal to a delay a solution belongs to $C[-\tau, 0]$.

In mathematics and its applications an important role is to use convenient notation. The convenient notation can emphasize essential features and shadow nonessential details of problems.

In the theory of ordinary differential equations one of such convenient notation is the *conditional representation* of ODE. For example, for the equation (13.2) the conditional representation is

$$\dot{x} = g(t, x)\,, \tag{13.13}$$

i.e. the argument t is omitted in variable $x(\cdot)$.

Let us introduce the conditional representation for system (13.1). In the phase space H an element of trajectory of the system (13.1) is presented as the pair $\{x(t);\, x(t+s),\ -\tau \leq s < 0\} \in H$. Denoting $\{x, y(\cdot)\}_t = \{x(t);\, x(t+s), -\tau \leq s < 0\} \in H$ we obtain the following conditional representation of the system (13.1) in the phase space H

$$\dot{x} = f(t, x, y(\cdot))\,, \ \{x, y(\cdot)\} \in H\,. \tag{13.14}$$

Therefore, to obtain the conditional representation of a FDE necessary in the original equation substitute the *finite dimensional* component $x(t)$ by x and replace the *functional* component $x(t+s)$, $-\tau \leq s < 0$ by $y(s)$, $-\tau \leq s < 0$. Note in (13.14) different symbols are used different symbols

$\{x, y(\cdot)\}_t$ for denoting $x(t) = x$ and function pre-history $x(t + \cdot) = y(\cdot)$, because they play different roles in the dynamic of systems with delays.

For systems (13.3) and (13.4) the conditional representation has the form

$$\dot{x} = G[t, x, y(-\tau(t))]\,; \tag{13.15}$$

$$\dot{x} = \int_{-\tau}^{0} \Phi[t, s, y(s)]\, ds\,. \tag{13.16}$$

In particularly, the conditional representation of the equation

$$\dot{x}(t) = bx(t) - \int_{-\tau}^{0} (\tau + s)x(t + s)ds$$

is

$$\dot{x} = bx - \int_{-\tau}^{0} (\tau + s)y(s)ds\,.$$

Note, the conditional representation (13.14) has no "physical sense" (as well as the conditional representation (13.13) in case of ODE), under (13.14) necessary to understand system (13.6), cosidering in the phase space H.

The importance of introducing the phase space H and the conditional representation (13.14) consists of possibility of explicitly separating the current – finite dimensional point – x and functional delay $y(\cdot)$.

Utilization of the space H and the conditional representation (13.14) allows us separate in the structure and behaviour of FDEs finite-dimensional and infinite-dimensional properties, and to introduce the system of notation which is the analog of ODE notation. These allows us to formulate the results in such a way, that when the delay disappears then all results coincide, with the exactness in notation, and with the corresponding results of the ODE Theory.

14 Existence and uniqueness of FDE solutions

Consider several definitions and propositions of the functional analysis.

Let Π be a Banach (i.e. complete linear normed) space. Operator $T : \Pi \to \Pi$ is *completely continuous* operator, if it is continuous and transfers every bounded set into a pre-compact set. The following classic proposition is valid

Lemma 14.1 *(Fix point principle). Let Ω be a closed bounded convex set of a Banach space Π and $T\colon \Omega \to \Omega$ is a pre-compact operator, then the operator has a fix point, i.e. there exists an element $Y \in \Omega$ such that $T(Y) = Y$.*

Let D be a subset of the space $C[a, b]$ continuous on $[a, b]$ n-dimensional functions. The set D is called *uniformly continuous*, if for any $\varepsilon > 0$ there exists $\delta > 0$ such that for any $Y(\cdot) \in D$, $t_1 \in [a, b]$, $t_2 \in [a, b]$, from the condition $|t_1 - t_2| < \delta$ follows $\|Y(t_1) - Y(t_2)\| < \varepsilon$.

The set D is called *uniformly bounded*, if there exists a number C such that for every $Y(\cdot) \in D$, $t \in [a, b]$ we have $\|Y(t)\| < C$.

The following proposition is valid

Lemma 14.2 *(Compactness Criterium). A closed set D is compact in $C[a, b]$, if and only if it is uniformly bounded and uniformly continuous.*

14.1 The classic solutions

Piece-wise continuous initial functions.

In this section we consider the solvability of the *Cauchy problem* (the Initial Value Problem)

$$\dot{x}(t) = f(t, x(t), x_t(\cdot))\,, \tag{14.1}$$

$$f(\cdot, \cdot, \cdot) : \mathbb{R} \times \mathbb{R}^n \times Q[-\tau, 0) \to \mathbb{R}^n\,,$$

in the space H.

Definition 14.1 *Solution $x(t; t_0, h^0)$ of the system (14.1) corresponding to an initial condition: $t_0 \in \mathbb{R}$ and $h^0 = \{x^0, y^0(\cdot)\} \in H$ is the function $x(\cdot)$, defined on an interval $[t_0 - \tau, t_0 + \kappa)$ ($\kappa > 0$), continuous and piece-wise differentiable on $[t_0, t_0 + \kappa)$, satisfying the equation (14.1) on $[t_0, t_0 + \kappa)$ (at points of discontinuities of derivatives in the equation (14.1) under $\dot{x}$ understand the right-hand side derivative) and the initial conditions:*

$$x(t_0) = x^0\,; \quad x(t_0 + s) = y^0(s) \ \textit{for} \ -\tau \leq s < 0\,.$$

Denote $Y_t \equiv \{Y(t); Y(t+s), \ -\tau \leq s < 0\}$ for $Y(\cdot) \in$ $\in E[h]$. Hence $x_t(t_0, h^0) = \{x(t; t_0, h^0); x(t+s; t_0, h^0), \ -$ $\tau \leq\leq s < 0\}$ is the presentation of the solution $x(t; t_0, h^0)$ in the space H. Note that x_t and $x_t(\cdot)$ have different meaning:

$$x_t(\cdot) \equiv \{x(t+s), \ -\tau \leq s < 0\} \in Q[-\tau, 0)\,,$$

$$x_t \equiv \{x(t); x(t+s), \ -\tau \leq s < 0\} \in \mathbb{R}^n \times Q[-\tau, 0)\,.$$

Definition 14.2 *Mapping $f(t, h) : \mathbb{R} \times H \to \mathbb{R}^n$ is locally Lipschits with respect to $h = \{x, y(\cdot)\} \in H$, if for any bounded set $D \subset H$ there exists a constant $L > 0$ such that*

$$\|f(t, h^{(1)}) - f(t, h^{(2)})\| \leq L\|h^{(1)} - h^{(2)}\|_H$$

for $t \in \mathbb{R}$ and $h^{(1)}$, $h^{(2)} \in D$.

Definition 14.3 *Mapping $f(t, h) : \mathbb{R} \times H \to \mathbb{R}^n$ is continuous on shift if for any $(\zeta, h) \in \mathbb{R} \times H$ there exists $\kappa > 0$ such that for all $Y(\cdot) \in E_\kappa[h]$ the corresponding function $\phi_Y(t) = f(t, Y_{t-\zeta})$ is continuous on the interval $[\zeta, \zeta + \kappa]$.*

Theorem 14.1 *Let the mapping $f(t, h)$ in system (14.1) be:*

1) locally Lipschits with respect to $h = \{x, y(\cdot)\}$;

2) continuous on shift.

Then for any $(t_0, h^0) \in \mathbb{R} \times H$ there exists $\bar{\kappa} > 0$ such that the unique solution $x(\cdot; t_0, h^0)$ of the system (14.1) is defined on $[t_0 - \tau, t_0 + \bar{\kappa}]$.

Proof. First we prove the existence of a solution for an arbitrary initial pair $h^0 = \{x^0, y^0(\cdot)\} \in H$.

Due to the first condition of the theorem, for $\{x^0, y^0(\cdot)\}$ there exists $\kappa > 0$ such that for any function $Y(\cdot) \in E_\kappa[x^0, y^0(\cdot)]$ the function–composition $\Psi_Y(\xi) = f(\xi, Y(\xi - t_0), Y_{\xi - t_0}(\cdot))$ is continuous on $[t_0, t_0 + \kappa]$.

For $\beta > 0$ consider the set $E_\kappa^\beta[x^0, y^0(\cdot)]$ of functions $Y(\cdot) \in E_\kappa[x^0, y^0(\cdot)]$ such that

$$\|Y(s) - x^0\| \leq \beta\,, \quad 0 < s \leq \kappa.$$

In the space $X = E_\kappa[x^0, y^0(\cdot)]$ introduce the operation of summation

$$Y^1(\cdot) + Y^2(\cdot) = \begin{cases} y^0(s), & -\tau \leq s < 0, \\ x^0, & s = 0, \\ Y^1(s) + Y^2(s) - x^0, & 0 < s \leq \kappa \end{cases}$$

and multiplication by a number $\mu \in \mathbb{R}$

$$\mu Y(\cdot) = \begin{cases} y^0(s), & -\tau \leq s < 0, \\ x^0, & s = 0, \\ \mu\,(Y(s) - x^0) + x^0, & 0 < s \leq \kappa, \end{cases}$$

and the norm

$$\|Y(\cdot)\|_X = \max_{0 \leq s \leq \kappa} \|Y(s) - x^0\|.$$

One can verify that X is the Banach space, moreover, the set $E_\kappa^\beta[x^0, y^0(\cdot)]$ is the bounded, closed, convex set in X.

On the set $E_\kappa^\beta[x^0, y^0(\cdot)]$ define the operator P:

$$T(Y(\cdot)) = \begin{cases} y^0(s), \; s \in [-\tau, 0), \\ x^0 + \int_{t_0}^{t_0+s} f(t, Y(t - t_0), Y_{t-t_0}(\cdot))dt, \; s \in [0, \kappa]. \end{cases}$$

Decreasing κ we can obtain that the operator T transfers the set $E^{\beta}_{\kappa}[x^0, y^0(\cdot)]$ into itself.

Define the function $\hat{Y}(\cdot) \in E^{\beta}_{\kappa}[x^0, y^0(\cdot)]$ by the rule:

$$\hat{Y}(\cdot) = \begin{cases} y^0(s), \ s \in [-\tau, 0), \\ x^0, \ s \in [0, \kappa]. \end{cases}$$

the function $f(\xi, \hat{Y}(\xi - t_0), \hat{Y}_{\xi - t_0}(\cdot))$ is continuous along $\hat{Y}(\cdot)$, and hence is bounded on $[0, \kappa]$, i.e. there is a number K, such that

$$\|f(\xi, \hat{Y}(\xi - t_0), \hat{Y}_{\xi - t_0}(\cdot))\| \leq K, \quad \xi \in [0, \kappa].$$

Then

$$\max_{0 \leq s \leq \kappa} \|Y(s) - x^0\| =$$

$$= \max_{0 \leq s \leq \kappa} \|x^0 + \int_{t_0}^{t_0+s} f(t, Y(t - t_0), Y_{t-t_0}(\cdot))dt - x^0\| \leq$$

$$\leq \max_{0 \leq s \leq \kappa} \| \int_{t_0}^{t_0+s} (f(t, Y(t - t_0), Y_{t-t_0}(\cdot))dt - f(t, \hat{Y}(t - t_0), \hat{Y}_{t-t_0}(\cdot)))dt\| +$$

$$+ \max_{0 \leq s \leq \kappa} \| \int_{t_0}^{t_0+s} f(t, \hat{Y}(t - t_0), \hat{Y}_{t-t_0}(\cdot))dt\| \leq (L + M)\beta\kappa + K\kappa$$

(here we applied the condition 2 of the theorem).

Take

$$\hat{\kappa} = \min \left\{ \kappa, \frac{\beta}{(L + M)\beta + K} \right\},$$

then

$$\max_{0 \leq s \leq \kappa} \|Y(s) - x^0\| \leq \beta,$$

that means $T(E^{\beta}_{\hat{\kappa}}[x^0, y^0(\cdot)]) \subset E^{\beta}_{\hat{\kappa}}[x^0, y^0(\cdot)]$.

Denote $D \equiv T(E^{\beta}_{\hat{\kappa}}[x^0, y^0(\cdot)])$ and prove that the set D is the uniformly continuous and uniformly bounded set in the space $C[0, \hat{\kappa}]$. Because $D \subset E^{\beta}_{\hat{\kappa}}[x^0, y^0(\cdot)]$, then for all $Y \in D$, $t \in [0, \hat{\kappa}]$ we have $\|Y(t)\| \leq \|x^0\| + \beta$, that means the uniform boundedness of D.

Let us prove the uniform continuity of D. Fix the arbitrary $Y \in D$ and $t_1, t_2 \in [0, \hat{\kappa}]$. By the definition of the set $D = T(E^{\beta}_{\hat{\kappa}}[x^0, y^0(\cdot)])$ there exists $\check{Y} \in E^{\beta}_{\hat{\kappa}}[x^0, y^0(\cdot)]$ such that $T(\check{Y}) = Y$.

Then

$$\|Y(t_1) - Y(t_2)\| =$$

$$= \|x^0 + \int_{t_0}^{t_0+t_1} f(t, \check{Y}(t-t_0), \check{Y}_{t-t_0}(\cdot))dt - x^0 -$$

$$- \int_{t_0}^{t_0+t_2} f(t, \check{Y}(t-t_0), \check{Y}_{t-t_0}(\cdot))dt\| \leq$$

$$\leq \int_{t_0+t_1}^{t_0+t_2} \|f(t, \check{Y}(t-t_0), \check{Y}_{t-t_0}(\cdot)) - f(t, \hat{Y}(t-t_0), \hat{Y}_{t-t_0}(\cdot))\|dt +$$

$$+ \int_{t_0+t_1}^{t_0+t_2} \|f(t, \hat{Y}(t-t_0), \hat{Y}_{t-t_0}(\cdot))\|dt \leq$$

$$\leq (L+M)\beta|t_1 - t_2| + K|t_1 - t_2|,$$

that means the uniform continuity of the set D.

It is easy to verify that the set D is closed. Then by the the lemma 14.2 D is the compact set. Therefore P is the pre-compact operator and according to the fix point theorem (lemma 14.1) there exists a fixed point, i.e. there exists $Y(\cdot) \in D$, such that

$$Y(s) = x^0 + \int_{t_0}^{t_0+s} f(t, Y(t-t_0), Y_{t-t_0}(\cdot))dt, \; s \in [0, \hat{\kappa}].$$

This equality is equivalent to the fact that $Y(\cdot)$ is the solution of the system (14.1).

Now prove the uniqueness. Let us assume the opposite, i.e. suppose that on $[t_0, t_0 + \hat{\kappa}]$ two solutions: $x^{(1)}(t)$ and $x^{(2)}(t)$ are defined. Without a loss of generality one can consider that $\hat{\kappa}$ satisfies the condition

$$(L+M)\hat{\kappa} < 1.$$

From the continuity of $x^{(1)}(t)$ and $x^{(2)}(t)$ follows existence of a moment $t_1 \in [t_0, t_0 + \hat{\kappa}]$ such that

$$\max_{t \in [t_0, t_0+\hat{\kappa}]} \|x^{(1)}(t) - x^{(2)}(t)\| = \|x^{(1)}(t_1) - x^{(2)}(t_1)\|.$$

Then

$$\|x^{(1)}(t_1) - x^{(2)}(t_1)\| =$$

$$= \|x^0 + \int_{t_0}^{t_1} f(t, x^{(1)}(t), x_t^{(1)}(\cdot))dt - x^0 - \int_{t_0}^{t_1} f(t, x^{(2)}(t), x_t^{(2)}(\cdot))dt\| \leq$$

$$\leq (L+M)\hat{\kappa}\|x^{(1)}(t_1) - x^{(2)}(t_1)\| < \|x^{(1)}(t_1) - x^{(2)}(t_1)\|,$$

that is possible only if $x^{(1)}(t_1) = x^{(2)}(t_2)$. This means that $x^{(1)}(t) = x^{(2)}(t)$ for all $t \in [t_0, t_0 + \hat{\kappa}]$. □

Remark 14.1 To prove the existence (but not uniqueness) of a solution of the system (14.1) the condition 1 of the theorem 14.1 can be replaced by the following relaxed condition:

1′) *Mapping f transfers bounded sets from H into bounded sets in $\mathbb{R}^n$.*

For investigating functional differential equations in the space H the condition 2 (shift continuity of the mapping f) of the Theorem 14.1 generally speaking cannot be replaced by the following[18]:

2′) *Mapping $f(t,h)$ is continuous with respect to t and h.*

because for a (*discontinuous*) pair $h = \{x, y(\cdot)\} \in H$ and any $Y(\cdot) \in E_\kappa[h]$, $\kappa > 0$, the mapping $Y_\xi : [0,\kappa] \to H$ has discontinuities at every points of the interval $[0,\kappa]$[19], so the conditions 1′ and 2′ do not guaranty for the composition $f(t, Y_{t-\xi})$ the smoothness necessary for proving the theorem 14.1.

Example 14.1 Let a function $h^0 = \{x^0, y^0(\cdot)\} \in H$ have discontinuity at a point $-\tau_0 \in (-\tau, 0]$. Fix a $\kappa \in (0, \tau - \tau_0)$. The interval $[0,\kappa]$ contains a nonmeasurable set $A \subset [0,\kappa]$. Consider a set

$$Z \equiv \left\{ \bigcup Y_t : Y(\cdot) \in E_\kappa[h^0],\ t \in [0,\kappa] \right\} \subset H$$

[18] which is the analog of ODE case

[19] see example 14.1

and define on it the mapping $f^0(\cdot) : Z \to \mathbb{R}^n$ by the rule

$$f^0(Y_t) = \begin{cases} x_1 & t \in A\,, \\ x_2 & t \in [0,\kappa]\backslash A\,, \end{cases}$$

where x_1, $x_2 \in \mathbb{R}^n$, moreover, $x_1 \neq x_2$. The set Z is closed in H, and the mapping f^0 is continuous on Z. Hence according to Tytse-Uryson theorem, the mapping f^0 can in continuous manner be extended from Z on all H.

Then taking into account that A and $[0,\kappa]\backslash A$ are non-measurable sets and hence for any $Y(\cdot) \in E_\kappa[h^0]$ the corresponding composition $f^0(Y_t)$ is non-measurable on $[0,\kappa]$, obtain that equation $\dot{x} = f^0(x, y(\cdot))$ does not have solutions for the initial function h^0 not in the sense of the definition 14.1, not even in the Caratheodory sense. □

For systems with discrete delays (13.15) the condition 2 of the theorem 14.1 is satisfied if the mapping G is continuous in domain. Also it is not difficult to verify that condition 2 of the theorem 14.1 is valid and for mappings $f(t,x,y(\cdot)) = P(t,x,\eta[y(\cdot)])$, where $P : \mathbb{R} \times \mathbb{R}^n \times \mathbb{R} \to \mathbb{R}^n$ are continuous functions and $\eta[y(\cdot)] : Q[-\tau, 0) \to \mathbb{R}$ is a regular functional, i.e. it can be presented in the following or more general integral form

$$\int_{-\tau}^{0} \alpha[s, y(s)]ds\,,$$

$$\int_{-\tau}^{0} \phi(u, \int_{u}^{0} \omega[s, y(s)]ds\,)du\,,$$

$$\int_{-\tau}^{0}\int_{-\tau}^{0} \gamma[s, u; y(s), y(s)]dsdu\,,$$

here $\alpha : [-\tau, 0] \times \mathbb{R}^n \to \mathbb{R}$, $\phi : [-\tau, 0] \times \mathbb{R} \to \mathbb{R}$, $\omega : [-\tau, 0] \times \mathbb{R}^n \to \mathbb{R}$, $\gamma : [-\tau, 0] \times [-\tau, 0] \times \mathbb{R}^n \times \mathbb{R}^n \to \mathbb{R}$ are continuous functions.

Remark 14.2 If in system (14.1) mapping $f(t, x, y(\cdot))$ has the support separated from the zero with respect to the functional variable $y(\cdot)$, i.e. there exists $\varepsilon \in (0, \tau)$ and a functional

$$F[t, x, y_\sigma(\cdot)] : \mathbb{R} \times \mathbb{R}^n \times Q_\sigma[-\tau, 0) \to \mathbb{R}^n$$

(here $\sigma = [-\varepsilon, 0)$) such that

$$f(t, x, y(\cdot)) = F[t, x, y_\sigma(\cdot)] \ ,$$

then existence of a theorem can be proved by step-by step advancement on the intervals of the length ε for sufficiently weak conditions on dependence of mapping $F[t, x, y_\sigma(\cdot)]$ on $y_\sigma(\cdot)$. In particular, this fact allows one effectively investigate equations with discrete delays .

Remark 14.3 In distinction from ODE solutions of FDEs, corresponding to different initial functions can intersect.

For example, for the equation

$$\dot{x}(t) = -x(t - \tau)$$

two solutions $x^{(1)}(t)$ and $x^{(2)}(t)$, corresponding to the following initial conditions:

1) $x(s) \equiv 1, \ -1 \leq s \leq 0,$

2) $x(s) \equiv -1, \ -1 \leq s \leq 0$

intersect at point $t = 1$: $x^{(1)}(1) = x^{(2)}(1) = 0$. However it does not contradict to existence and uniqueness theorem, because intersect only finite-dimensional components, and at this point pre-histories are different.

Remark 14.4 From theorem 14.1 it follows that there is a uniqueness of solutions in the positive direction of t. In distinction from ODE in the negative direction of t the theorem is not valid, i.e. there is a possible effect when the solution cannot be continued in negative direction of t.

Continuous initial functions

If we study the (14.1) in the class of continuous initial functions, then one can prove the existence and uniqueness conditions similar to ODE case.

Theorem 14.2 *Let mapping $f(t,h) : \mathbb{R} \times C[-\tau, 0] \to \mathbb{R}^n$:*

1) Locally Lipshits with respect to h in $\mathbb{R} \times C[-\tau, 0]$;

2) continuous on $\mathbb{R} \times C[-\tau, 0]$.

Then for any $(t_0, h^0) \in \mathbb{R} \times C[-\tau, 0]$ there exists $\bar{\kappa} > 0$ such that on $[t_0 - \tau, t_0 + \bar{\kappa}]$ the unique solution $x(\cdot; t_0, h^0)$ of system (14.1) is defined.

On the continuity of solutions and dependence on initial data

Further the maximal right interval of existence a solution $x(\cdot; t_0, h)$ will be denoted by $J^+(t_0, h)$ or just J^+.

Theorem 14.3 *Let the conditions of theorem 14.1 be fulfilled. Then for all $(t_0, h) \in \mathbb{R} \times H$ either $J^+(t_0, h) = [t_0, +\infty)$, or $J^+(t_0, h) = [t_0, +\beta)$ $(t_0 < \beta < +\infty)$ and $\lim\limits_{t \to \beta - 0} x(t; t_0, h) = \infty$.*

Theorem 14.4 *Let the conditions of theorem 14.1 are fulfilled. Then $(\forall \gamma > 0)$ $(\exists L > 0)$ $(\forall t_0 \in \mathbb{R})$ $(\forall h_1, h_2 \in B_\gamma = \{x \in \mathbb{R}^n : \|x\| < \gamma\})$ for all $t \geq t_0$, for which segments of solutions $x_t(t_0, h_1)$, $x_t(t_0, h_1)$ belongs to region B_γ, the following inequality is valid*

$$\|x(t; t_0, h_1) - x(t; t_0, h_2)\| \leq \|h_1 - h_2\|_H \, e^{L\,(t-t_0)} \,. \quad (14.2)$$

14.2 Caratheodory solutions

If in the system

$$\dot{x}(t) = f(t, x(t), x_t(\cdot)) \quad (14.3)$$

the mapping

$$f(\cdot,\cdot,\cdot) : \mathbb{R} \times \mathbb{R}^n \times Q[-\tau, 0) \to \mathbb{R}^n$$

does not satisfy conditions of the theorem 14.1, then the notion of the solution can be extended. For simplicity we consider only the cases of *continuous* initial functions.

Definition 14.4 *Absolutely continuous solution (the solution in Caratheodory sense) $x(t; t_0, h^0)$ of the system (14.3) with initial conditions $t_0 \in \mathbb{R}$ and $h^0 = \{x^0, y^0(\cdot)\} \in C[-\tau, 0]$ is a function $x(\cdot)$, defined on an interval $[t_0 - \tau, t_0 + \kappa)$ $(\kappa > 0)$, absolutely continuous on $[t_0, t_0 + \kappa)$, satisfying almost everywhere on $[t_0, t_0 + \kappa)$ the equation (14.3) and the initial conditions:*

$$x(t_0) = x^0\,; \quad x(t_0 + s) = y^0(s) \quad \textit{for} \ -\tau \le s < 0\,.$$

Consider under what restrictions on mapping f one can prove the existence of solutions in the Caratheodory sense.

Definition 14.5 *Mapping $f(t, h) : \mathbb{R} \times C[-\tau, 0] \to \mathbb{R}^n$ satisfies the Caratheodory conditions, if:*

1) *f is measurable with respect to t for every fixed $h \in C[-\tau, 0]$,*

2) *f is continuous with respect to $h \in C[-\tau, 0]$ for every fixed t,*

3) *For any point $(t, h) \in \mathbb{R} \times C[-\tau, 0]$ there exists a neighbourhood $O_{t,h}$ and integrable by Lebesgue function $k(\xi) \ge 0$, $\xi \ge t$, such that*

$$\|f(\xi, h)\| \le k(\xi)\,, \quad (\xi, h) \in O_{t,h}\,, \tag{14.4}$$

$$\|f(\xi, h_1) - f(\xi, h_2)\| \le k(\xi)\ \|h_1 - h_2\|_C\,, \\ (\xi, h_1) \& (\xi, h_2) \in O_{t,h}\,. \tag{14.5}$$

The following proposition is valid

Theorem 14.5 *Let mapping $f(t,h)$ in the system (14.3) satisfies the Caratheodory conditions. Then for any $(t_0,h^0) \in \mathbb{R} \times C[-\tau,0]$ there exists $\bar{\kappa} \in (0,\kappa]$ such that on $[t_0-\tau, t_0+\bar{\kappa}]$ the unique absolute continuous solution $x(\cdot;t_0,h^0)$ of system (14.3) is defined, the solution continuously depends on h^0.*

14.3 The step method for systems with discrete delays

One of the most simple methods of solving equations with discrete delays (13.3) is the *method of steps* (method of successive integration). For simplicity we consider the case of the constant discrete delay: $\tau(t) = \tau$). Let $t_0 \in \mathbb{R}$ be an initial moment and $h^0 = \{x^0, y^0(\cdot)\} \in H$ an initial function.

Then on the interval $[t_0, t_0+\tau]$ the solution of the system (13.3) coincides with the solution of the system of *ordinary* differential equations:

$$\begin{cases} \dot{x}(t) = G[t, x(t), y^0(t-\tau)]\,, \\ x(t_0) = x^0\,. \end{cases} \tag{14.6}$$

The system (14.6) is obtained from (13.3) by substituting the initial function $y^0(\cdot)$ into (13.3). This is the first step.

If the solution $x(t) = \phi_1(t)$ of the Cauchy problem (14.6) exists on the whole interval $[t_0, t_0+\tau]$, then the solution of system (13.3) on the next interval $[t_0+\tau, t_0+2\tau]$ coincides with the solution of the following ODE initial value problem:

$$\begin{cases} \dot{x}(t) = G[t, x(t), \phi_1(t-\tau)]\,, \\ x(t_0+\tau) = \phi_1(t_0+\tau)\,. \end{cases}$$

This is the second step.

Realising successively similar constructions one can find the solution of the system (13.3) on any bounded interval.

Practical application of this method is a laborious procedure, however it is useful for finding the exact solutions of simple equations which can be used for testing numerical methods.

15 Smoothness of solutions and expansion into the Taylor series

In this section we consider some aspects of solution smoothness of in the initial value problem:

$$\dot{x}(t) = f(t, x(t), x_t(\cdot)); \tag{15.1}$$

$$x(t_0) = x^0; \tag{15.2}$$

$$x(t_0 + s) = y^0(s), -\tau \leq s < 0. \tag{15.3}$$

Obviously smoothness of the initial function $\{x^0, y^0(\cdot)\}$ is generally necessary (but not a sufficient conditions) for smoothness of the corresponding FDE solution. For example, if we investigate continuity of $(k + 1)$-th derivative of the solution of the system

$$\dot{x}(t) = G[t, x(t), x(t - \tau)], \tag{15.4}$$

then the corresponding initial function $\{x^0, y^0(\cdot)\}$ should be k times the continuous differentiable. For this reason we always assume (without additional remarks) the necessary smoothness of initial functions, and we also assume fulfillment conditions, which guarantee the existence and uniqueness of solutions.

Smoothness of solutions at the initial moment

In order for the solutions of the problem (15.1) – (15.3) to be continuous differentiable at the initial moment t_0 necessary and sufficient to fulfillment the condition

$$\frac{dy^0(0)}{ds} = f(t_0, x^0, y^0(\cdot)), \tag{15.5}$$

which is called the *sewing condition of the first order* at the point $\{x^0, y^0(\cdot)\} \in C[-\tau, 0]$. For continuity of the m-th derivative of the solution at the initial moment t_0, it is necessary and sufficient to fulfillment the *sewing condition of the m-th order*

$$\frac{d^k y^0(0)}{ds^k} = \left(\frac{d^{k-1}}{dt^{k-1}} f(t_0, x^0, y^0(\cdot)) \right)_{(15.1)}, \quad k = 1, \ldots, m, \tag{15.6}$$

where $\dfrac{d^k y^0(0)}{ds^k}$ denotes k-th *right-hand side* derivative of the initial function at the zero, and

$$\left(\frac{d^{k-1}}{dt^{k-1}} f(t_0, x^0, y^0(\cdot)) \right)_{(15.1)}$$

is the k-th total derivative of the mapping $f(t, x, y(\cdot))$ with respect to the system (15.1) at the point $(t_0, x^0, y^0(\cdot))$.

The following proposition is valid.

Theorem 15.1 *Let the mapping $f(t, x, y(\cdot))$ has invariantly differentiable partial and invariant derivatives with respect to t, x, $y(\cdot)$ up to an order $m > 0$ in a neighborhood of a point $(t_0, x^0, y^0(\cdot))$. If the initial data t_0, x^0, $y^0(\cdot)$ satisfy the sewing condition (15.6), then the corresponding solution of the problem (15.1) – (15.3) is $(m+1)$ times continuous differentiable at the initial moment t_0.*

Smoothness of solutions on an interval

The following proposition gives the conditions of continuity of the solution of the problem (15.1) – (15.3) on the whole interval $[t_0 - \tau, t_0 + \theta]$.

Theorem 15.2 *Let the mapping $f(t, x, y(\cdot))$ has invariantly continuous partial and invariant derivatives with respect to t, x, $y(\cdot)$ up to an order $m > 0$ in the domain.*

If the initial data t_0, x^0, $y^0(\cdot)$ satisfy the sewing condition (15.6) and the corresponding solution $x(t)$ of problem (15.1) – (15.3) is defined on an interval $[t_0 - \tau, t_0 + \theta]$, then the solution is continuously differentiable up to the order $(m+1)$ on the interval $[t_0 - \tau, t_0 + \theta]$.

Proof. Because $x(t)$ is the solution of the problem (15.1) – (15.3), hence the function $x(t)$ satisfies the system of differential equations

$$\dot{x}(t) = f(t, x(t), x_t(\cdot)) . \tag{15.7}$$

Because $x(t)$ is continuous and $f(t, x, y(\cdot))$ is invariantly continuous, hence the composition $f(t, x(t), x_t(\cdot))$ is the continuous on $[t_0, t_0 + \theta]$ function. From this fact follows the continuity of the first derivative $\dot{x}(t)$ on the interval $[t_0, t_0 + \theta]$.

For $m > 1$ the mapping $f(t, x, y(\cdot))$ has invariantly continuous detrivatives $\frac{\partial f}{\partial t}, \frac{\partial f}{\partial x}$ and $\partial_y f$. So the right part of the system (15.7) is continuous differentiable with respect to t on the interval $[t_0, t_0+\theta]$. Hence, the *second* derivative of the solution $x(t)$ is contunuous on $[t_0, t_0 + \theta]$. Differentiating both parts of the equation (15.7) by t obtain

$$\ddot{x}(t) = \frac{\partial f(t, x(t), x_t(\cdot))}{\partial t} + \frac{\partial f'(t, x(t), x_t(\cdot))}{\partial x} f(t, x(t), x_t(\cdot)) +$$

$$+ \partial_y f(t, x(t), x_t(\cdot)) . \tag{15.8}$$

If $m > 2$, then the right-hand side of the system (15.8) is continuously differentiable by t on the interval $[t_0, t_0+\theta]$. So the third derivative of the solution $x(t)$ is to be continuous on $[t_0, t_0 + \theta]$.

Repeating the corresponding procedure m times we prove the theorem. □

Note, continuous differentiability of solutions at the initial moment follows from the theorem 15.1, because the corresponding sewing conditions are satisfied. Also it is necessary to note that in the framework of the theorem the necessary smoothness of the initial function is also assumed.

15.1 Density of special initial functions

In many cases (as an example for numerical methods) it is necessary that the exact solution would be smooth on the whole interval of existence $[t_0 - \tau, t_0 + \theta]$.

However in most of cases, the solutions are smooth only under fulfillment of sewing conditions (15.6), that is the sufficiently restricted assumption for real problems. If, for example sewing conditions are not setisfied, then in numerical procedures it is necessary to construct special algorithms in neighborhoods of discontinuity points.

In this section we show, that for many classes of FDEs sets of functions, satisfying sewing conditions, it is dense in $C[-\tau, 0]$. Therefore in many real problems one can as a rule assume that for a given initial function the sewing conditions are satisfied.

To simplify notation we consider further the *one-dimensional* equation (15.4) with constant delay. For the system (15.4) and an initial function $\{x^0, y^0(\cdot)\} \in C[-\tau, 0]$ the sewing condition of the first order has the form

$$\frac{dy^0(0)}{ds} = G[t_0, x^0, y^0(-\tau)]\,. \tag{15.9}$$

The sewing condition of the second order is

$$\frac{d^2y^0(0)}{ds^2} = \frac{d}{dt}G[t_0, x^0, y^0(-\tau)] =$$

$$= \partial_t G[t_0, x^0, y^0(-\tau)] + \frac{\partial G[t_0, x^0, y^0(-\tau)]}{\partial x}\, G[t_0, x^0, y^0(-\tau)]\,. \tag{15.10}$$

The sewing condition of the third order has the form:

$$\frac{d^3y^0(0)}{ds^3} = \partial_t^2 G[t_0, x^0, y^0(-\tau)] +$$

$$+\left[\frac{\partial}{\partial x}\left(\partial_t G[t_0, x^0, y^0(-\tau)]\right)\right] G[t_0, x^0, y^0(-\tau)] +$$

$$+\left[\partial_t\left(\frac{\partial G[t_0, x^0, y^0(-\tau)]}{\partial x}\right)\right] G[t_0, x^0, y^0(-\tau)] +$$

$$+\frac{\partial^2 G[t_0, x^0, y^0(-\tau)]}{\partial x^2} G^2[t_0, x^0, y^0(-\tau)] +$$

$$+\frac{\partial G[t_0, x^0, y^0(-\tau)]}{\partial x} \partial_t G^2[t_0, x^0, y^0(-\tau)] +$$

$$+\left(\frac{\partial G[t_0, x^0, y^0(-\tau)]}{\partial x}\right)^2 G^2[t_0, x^0, y^0(-\tau)]. \qquad (15.11)$$

Consider an initial function $h_* = \{x^*, y^*(\cdot)\} \in C[-\tau, 0]$, which does not satisfy sewing conditions, and fix a $\Delta \in (0, \tau)$.

Right-hand sides of (15.9) – (15.11) do not depend on values the function $y^*(\cdot)$ on subinterval $(-\Delta, 0)$. Hence the function $y^*(\cdot)$ can be changed (modified) on the interval $(-\Delta, 0)$ in a such manner, that the initial function $h_\Delta = \{x^*, y^*_\Delta(\cdot)\} \in C[-\tau, 0]$, corresponding to the modified function $y^*_\Delta(\cdot)$, will satisfy the sewing conditions of a required order. Moreover, the norm of difference $\|h_* - h_\Delta\|_C$ can be made arbitrarily small.

Because initial functions h_* and h_Δ are closed, the corresponding solutions also will be closed (due to continuous dependence of FDE solutions on initial data). So for sufficiently wide classes of FDEs for realization of numerical methods one can assume that a (given) initial function satisfies the sewing conditions if Δ is chosen sufficiently small (less than a step of numerical integration).

In the general case the following remark is valid

Remark 15.1 If mapping $G[\cdot, \cdot, \cdot]$ in system (15.4) m times continuous differentiable with respect to all variables, then the set of functions satisfying the sewing conditions up to order $(m + 1)$ is dense in $C[-\tau, 0]$.

A similar remark is valid for general FDEs with supports, separated from the zero.

Remark 15.2 If mapping $f(t, x, y(\cdot))$ has support with respect to $y(\cdot)$, separated from the zero, and *invariantly continuous* partial and invariant derivatives (including mixed derivatives) with respect to t, x and $y(\cdot)$ up to an order $m > 0$ then the set of functions $\phi(\cdot) \in C^m[-\tau, 0]$, satisfying *sewing condition* (15.6) is dense in $C[-\tau, 0]$.

Remark 15.3 If a functional differential equation with *distributed* delays (of the integral type) has the support, separated from the zero, then one can construct similar to the discrete case the corresponding sets of continuous on $[-\tau, 0]$ initial functions, which satisfy the sewing conditions (15.6). As a rule, such classes of functions are dense in $C[-\tau, 0]$.

15.2 Expansion of FDE solutions into Taylor series

For the sake of further notation simplicity we consider the one-dimensional equation (15.1).

Suppose that $x(t)$ is a sufficiently smooth solution of the equation (15.1). Consider the expansion of the solution into the Taylor series in a right-hand neighborhood of the point $t_* \in [t_0, \theta)$:

$$x(t) = x_* + x_*^{(1)}(t - t_*) + x_*^{(2)}\frac{(t-t_*)^2}{2!} + x_*^{(3)}\frac{(t-t_*)^3}{3!} + \ldots =$$

$$= \sum_{k=0}^{\infty} \frac{x_*^{(k)}}{k!}(t - t_*)^k, \qquad (15.12)$$

where $x_*^{(k)} = x^{(k)}(t_*)$, $k = 1, 2, \ldots$, denotes k-th derivative of the solution $x(t)$ at point t_*.

Hence, for a natural N a finite sum

$$x_N(t) = \sum_{k=0}^{N} \frac{x_*^{(k)}}{k!}(t - t_*)^k \qquad (15.13)$$

is an approximation of the exact solution in an neighborhood of the point t_* (if one can calculate the derivatives $x_*^{(k)} = x^{(k)}(t_*)$, $k = 1, \ldots, N$).

Describe the procedure of calculating derivatives $x_*^{(k)}$. We use the following notation: $\frac{\partial^m f(t, x, y(\cdot))}{\partial x^m}$ is an m-th partial derivative with respect to x, and $\partial_t^m f(t, x, y(\cdot))$ is the coinvariant derivative.

Further we assume the existence of all necessary derivatives.

1) Substituting $t = t_*$ into (15.1), we obtain the formula for finding $x_*^{(1)} = x^{(1)}(t_*)$:

$$x_*^{(1)} = \frac{dx(t_*)}{dt} = f(t_*, x(t_*), x_{t_*}(\cdot)) \,. \qquad (15.14)$$

2) To find $x_*^{(2)} = x^{(2)}(t_*)$ differentiate both sides of (15.1):

$$\begin{aligned} x_*^{(2)} &= \left.\frac{d^2x(t)}{dt^2}\right|_{t=t_*} = \frac{d}{dt}\left(\frac{dx(t)}{dt}\right)\Big|_{t=t_*} = \\ &= \left(\frac{d}{dt} f(t, x(t), x_t(\cdot))\right)_{t=t_*} = \\ &= \partial_t f(t_*, x(t_*), x_{t_*}(\cdot)) + \\ &+ \frac{\partial f(t_*, x(t_*), x_{t_*}(\cdot))}{\partial x} f(t_*, x(t_*), x_{t_*}(\cdot)) \,. \end{aligned} \qquad (15.15)$$

3) To find $x_*^{(3)} = x^{(3)}(t_*)$ differentiate both sides of (15.15)

$$x_*^{(3)} = \left.\frac{d^3x(t)}{dt^3}\right|_{t=t_*} = \frac{d}{dt}\left(\frac{d^2x(t)}{dt^2}\right)\Big|_{t=t_*} =$$

$$= \Bigg(\partial_t^2 f(t, x(t), x_t(\cdot)) +$$

$$+ \left[\frac{\partial}{\partial x} \partial_t f(t, x(t), x_t(\cdot)) \right] f(t, x(t), x_t(\cdot)) +$$

$$+ \left[\partial_t \left(\frac{\partial f(t, x(t), x_t(\cdot))}{\partial x} \right) \right] f(t, x(t), x_t(\cdot)) +$$

$$+ \frac{\partial^2 f(t, x(t), x_t(\cdot))}{\partial x^2} f^2(t, x(t), x_t(\cdot)) +$$

$$+ \frac{\partial f(t, x(t), x_t(\cdot))}{\partial x} \partial_t f(t, x(t), x_t(\cdot)) +$$

$$+ \left(\frac{\partial f(t, x(t), x_t(\cdot))}{\partial x} \right)^2 f(t, x(t), x_t(\cdot)) \Bigg)_{t=t_*} =$$

$$= \partial_t^2 f(t_*, x(t_*), x_{t_*}(\cdot)) +$$

$$+ \left[\frac{\partial}{\partial x} \partial_t f(t_*, x(t_*), x_{t_*}(\cdot)) \right] f(t_*, x(t_*), x_{t_*}(\cdot)) +$$

$$+ \left[\partial_t \left(\frac{\partial f(t_*, x(t_*), x_{t_*}(\cdot))}{\partial x} \right) \right] f(t_*, x(t_*), x_{t_*}(\cdot)) +$$

$$+ \frac{\partial^2 f(t_*, x(t_*), x_{t_*}(\cdot))}{\partial x^2} f^2(t_*, x(t_*), x_{t_*}(\cdot)) +$$

$$+ \frac{\partial f(t_*, x(t_*), x_{t_*}(\cdot))}{\partial x} \partial_t f(t_*, x(t_*), x_{t_*}(\cdot)) +$$

$$+ \left(\frac{\partial f(t_*, x(t_*), x_{t_*}(\cdot))}{\partial x} \right)^2 f(t_*, x(t_*), x_{t_*}(\cdot)). \qquad (15.16)$$

4) Differentiating successively the corresponding relations one can calculate $x^{(4)}(t_*), x^{(5)}(t_*), \ldots, x^{(N)}(t_*)$.

In the general case the following formula is valid: $x^{(k)} = \left(\frac{d^{k-1} f(t,x,y(\cdot))}{dt^{k-1}} \right)_{(15.1)}$, where the right-hand side is the $(k-1)$ - the *total derivative* of the functional $f(t, x, y(\cdot))$ with respect to the system (15.1).

Remark 15.4 As described in this section, formulas have the same presentation as in ODE case. We just use (in the case of FDE) *coinvariant derivatives* with respect to t instead of the usual derivatives[20].

A finite sum of the Taylor series can be considered as a basis of a numerical method of approximate solving FDE. For a sufficiently smooth solution, the approximation error of (15.13) to the exact solution will be the value of the order $(t - t_*)^{N+1}$.

A positive feature of this numerical method consists of the following fact: (15.13) is expanded with respect to successive principal terms, that allow one without additional expenses (outlays) on the calculation of the remaining terms of the sum (15.13) to obtain an estimation of *local* error of a approximate solution.

However, such a one-step method of integrating FDEs requires at every step calculating $\frac{N(N+1)}{2}$ values of the functional f and its derivatives.

16 The sewing procedure

In the previous sections we showed that FDE solutionscould be k-times continuous differentiable, if the sewing conditions are valid, also it was shown that initial functions, which do not satisfy the sewing conditions, can be approximated by functions, satisfying the sewing conditions.

In this section we discuss the procedure of modifying an initial function in such a manner that it would satisfy the sewing conditions, i.e, we consider the following problem: Let an initial function not satisfy the sewing conditions. Is it possible a slightly change the function in such a manner that the modified function satisfies the sewing conditions. In the general case the answer is positive. Below we also discuss a possible modification procedure.

---

[20]ODE case

16.1 General case

For simplicity consider the *one-dimensional* equation

$$\dot{x}(t) = f(t, x(t), x_t(\cdot)) \tag{16.1}$$

with the initial condition

$$\begin{cases} x(t_0) = \phi(0)\,, \\ x(t_0 + s) = \phi(s)\,,\ -\tau \le s < 0\,, \end{cases} \tag{16.2}$$

where $\phi(\cdot) \in C^k[-\tau, 0]$.

In this section the m-th derivative of the function $\phi(\cdot)$ be denoting by $\phi^{(m)}(\cdot)$.

Our aim is to show that the initial function $\phi(\cdot)$ can be changed on an interval $[-\Delta, 0]$ (for some $\Delta \in [0, \tau)$) in such a manner that the modified initial function satisfies the sewing conditions of order k.

If the initial function $\phi(\cdot)$ satisfies the required sewing conditions, then we put $\Delta = 0$.

Let the opposite be valid, i.e. the original initial function does not satisfy the sewing conditions. Fix a $\Delta \in (0, \tau)$.

Consider the function

$$\psi_\Delta(\cdot) = \begin{cases} \phi(s)\,, -\tau \le s \le -\Delta\,, \\ \psi(s)\,, -\Delta \le s \le 0\,. \end{cases} \tag{16.3}$$

Obtained by modifying $\psi(s)$, $-\Delta \le s \le 0$ of the original function $\phi(\cdot)$ on the interval $[-\Delta, 0]$.

The function (16.3) will be k times continuously differentiable at the point $t = -\Delta$ if the following conditions are satisfied

$$\begin{cases} \psi(-\Delta) = \phi(-\Delta)\,, \\ \dot{\psi}(-\Delta) = \dot{\phi}(-\Delta)\,, \\ \psi^{(2)}(-\Delta) = \phi^{(2)}(-\Delta)\,, \\ \dots \\ \psi^{(k)}(-\Delta) = \phi^{(k)}(-\Delta)\,. \end{cases} \tag{16.4}$$

For the function (16.3) the sewing conditions (at point $t = 0$) have the form

$$\begin{cases} \psi(0) = \phi(0)\,, \\ \dot{\psi}(0) = f(t_0, \psi(0), \psi_\Delta(\cdot))\,, \\ \psi^{(2)}(0) = \left(\dfrac{df(t,x,y(\cdot))}{dt} \Bigg|_{(t_0,\psi(0),\psi_\Delta(\cdot))} \right)_{(16.1)}\,, \\ \dots \\ \psi^{(k)}(0) = \left(\dfrac{d^{k-1} f(t,x,y(\cdot))}{dt^{k-1}} \Bigg|_{(t_0,\psi(0),\psi_\Delta(\cdot))} \right)_{(16.1)}\,, \end{cases} \tag{16.5}$$

where

$$\left(\frac{d^m f(t,x,y(\cdot))}{dt^m} \Bigg|_{(t_0,\psi(0),\psi_\Delta(\cdot))} \right)_{(16.1)}$$

is the m-th order total derivative of the mapping $f(t, x, y(\cdot))$ with respect to the system (16.1) at the point $(t_0, \psi(0), \psi_\Delta(\cdot))$.

16.2 Sewing (modification) by polynomials

Take the function $\psi(\cdot)$ of the form

$$\psi(s) = \phi(0) + \sum_{i=1}^{k+1} \alpha_i\, s^i + \sum_{j=1}^{k} \beta_j\, s^{j+k+1}\,, \quad -\Delta \le s \le 0\,. \tag{16.6}$$

Substituting (16.6) into (16.4) obtain

$$\begin{pmatrix} \alpha_1(-\Delta) & +\alpha_2(-\Delta)^2 & +\alpha_3(-\Delta)^3 & +\alpha_4(-\Delta)^4 \\ \alpha_1 & +\alpha_2\,2\,(-\Delta) & +\alpha_3\,3\,(-\Delta)^2 & +\alpha_4\,4\,(-\Delta)^3 \\ 0 & +\alpha_2\,2 & +\alpha_3\,3\cdot 2\,(-\Delta) & +\alpha_4\,4\cdot 3\,(-\Delta)^2 \\ 0 & 0 & +\alpha_3\,3! & +\alpha_4\,4\cdot 3\cdot 2\,(-\Delta) \\ \cdots & \cdots & \cdots & \cdots \\ 0 & 0 & 0 & 0 \end{pmatrix} +$$

$$+ \begin{pmatrix} \alpha_5(-\Delta)^5 & + \ \ldots \ + \ \alpha_{k+1}(-\Delta)^{k+1} \\ \alpha_5\,5\,(-\Delta)^4 & + \ \ldots \ + \ \alpha_{k+1}\,(k+1)\,(-\Delta)^{k} \\ \alpha_5\,5\cdot 4\,(-\Delta)^3 & + \ \ldots \ + \ \alpha_{k+1}\,(k+1)\,k\,(-\Delta)^{k-1} \\ \alpha_5\,5\cdot 4\cdot 3\,(-\Delta)^2 & + \ \ldots \ + \ \alpha_{k+1}\,(k+1)\,k\,(k-1)\,(-\Delta)^{k-2} \\ \cdots & \cdots \quad \cdots \\ 0 & + \ \ldots \ + \ \alpha_k\,k! + \alpha_{k+1}\,\dfrac{(k+1)!}{1!}\,(-\Delta) \end{pmatrix} =$$

$$= \begin{pmatrix} \phi(-\Delta) - \phi(0) - \sum_{j=1}^{k} \beta_j(-\Delta)^{j+k+1} \\ \dot{\phi}(-\Delta) - \sum_{j=1}^{k} \beta_j\,(j+k+1)\,(-\Delta)^{j+k} \\ \phi^{(2)}(-\Delta) - \sum_{j=1}^{k} \beta_j\,(j+k+1)\,(j+k)\,(-\Delta)^{j+k-1} \\ \phi^{(3)}(-\Delta) - \sum_{j=1}^{k} \beta_j\,\dfrac{(j+k+1)!}{(j+k-2)!}\,(-\Delta)^{j+k-2} \\ \cdots \\ \phi^{(k)}(-\Delta) - \sum_{j=1}^{k} \beta_j\,\dfrac{(j+k+1)!}{(j+1)!}\,(-\Delta)^{j+1} \end{pmatrix} \tag{16.7}$$

System (16.7) is the linear system of $k+1$ linear algebraic equations with respect to $k+1$ unknown variables $\alpha_1,\ldots,\alpha_{k+1}$. The system has the unique solution

$$\begin{cases} \alpha_1 = g_1(\beta_1,\ldots,\beta_k)\,, \\ \ldots \\ \alpha_{k+1} = g_{k+1}(\beta_1,\ldots,\beta_k)\,, \end{cases} \tag{16.8}$$

where functions $g_i(\beta_1,\ldots,\beta_k)$, $i=1,\ldots,k+1$, are linear with respect to $\beta_1,\ldots,\beta_k$.

Substituting (16.8) into (16.5) we obtain the system of k algebraic equations with respect to $\beta_1,\ldots,\beta_k$ (note, we do not consider the first equation of (16.5)).

For the function(16.6)

$$\psi(0) = \phi(0)\,,\ \dot{\psi}(0) = \alpha_1\,,\ \psi^{(2)}(0) = \alpha_2\,,\ \ldots\,,\ \psi^{(k)}(0) = \alpha_k\,. \tag{16.9}$$

If the system has a solution $\beta_1^*,\ldots,\beta_k^*$, then substituting it into the (16.8) find $\alpha_i^* = g_i(\beta_1^*,\ldots,\beta_k^*)\,,\ i=1,\ldots,k+1$. The corresponding functions (16.6) give us the requiring modification.

16.3 The sewing procedure of the second order

As an example consider for the equation (16.1) the second order sewing procedure. The function (16.6) has the form

$$\psi(s) = \phi(0) + \alpha_1 s + \alpha_2\, s^2 + \alpha_3\, s^3 + \beta_1\, s^4 + \beta_2\, s^5\,,\ -\Delta \le s \le 0\,, \tag{16.10}$$

The corresponding system (16.7) has the representation

$$\begin{cases} \alpha_1(-\Delta) + \alpha_2(-\Delta)^2 + \alpha_3(-\Delta)^3 = \\ \qquad = \phi(-\Delta) - \phi(0) - \beta_1(-\Delta)^4 - \beta_2(-\Delta)^5\,, \\ \alpha_1 + \alpha_2\, 2\,(-\Delta) + \alpha_3\, 3\,(-\Delta)^2 = \\ \qquad = \dot{\phi}(-\Delta) - \beta_1\, 4\,(-\Delta)^3 - \beta_2\, 5\,(-\Delta)^4\,, \\ \alpha_2\, 2 + \alpha_3\, 6\,(-\Delta) = \\ \qquad = \phi^{(2)}(-\Delta) - \beta_1\, 12\,(-\Delta)^2 - \beta_2\, 20\,(-\Delta)^3\,. \end{cases}$$

One can find

$$\begin{cases} \alpha_1 = -\beta_1(-\Delta)^3 - \beta_2\, 3\,(-\Delta)^4 - \\ \qquad -\frac{3}{2}\frac{\phi(-\Delta)-\phi(0)}{\Delta} - 2\,\dot{\phi}(-\Delta) - -\frac{3}{2}\,\Delta\,\phi^{(2)}(-\Delta)\,, \\ \alpha_2 = \beta_1\, 3\,\Delta^2 + \beta_2\, 8\,\Delta^3 - \\ \qquad -2\,\frac{\phi(-\Delta)-\phi(0)}{\Delta^2} - 3\,\frac{\dot{\phi}(-\Delta)}{\Delta} - 2\,\phi^{(2)}(-\Delta)\,, \\ \alpha_3 = \beta_1\, 3\,\Delta - \beta_2\, 6\,\Delta^2 - \\ \qquad -\frac{1}{2}\frac{\phi(-\Delta)-\phi(0)}{\Delta^3} - \frac{\dot{\phi}(-\Delta)}{\Delta^2} - \frac{1}{2}\frac{\phi^{(2)}(-\Delta)}{\Delta}\,. \end{cases} \tag{16.11}$$

From (16.5) and (16.9) obtain two equations for β_1 and β_2:

$$\begin{cases} \alpha_1 = f(t_0, \phi(0), \psi_\Delta(\cdot))\,, \\ \alpha_2 = \left(\left. \frac{df(t,x,y(\cdot))}{dt} \right|_{(t_0,\psi(0),\psi_\Delta(\cdot))} \right)_{(16.1)}\,, \end{cases} \tag{16.12}$$

or in explicit form (taking into account (16.11))

$$\begin{cases} -\beta_1(-\Delta)^3 - \beta_2\, 3\,(-\Delta)^4 - \frac{3}{2}\,\frac{\phi(-\Delta)-\phi(0)}{\Delta} - \\ \qquad -2\,\dot{\phi}(-\Delta) - \frac{3}{2}\,\Delta\,\phi^{(2)}(-\Delta) = f(t_0, \phi(0), \psi_\Delta(\cdot))\,, \\ \beta_1\, 3\,\Delta^2 + \beta_2\, 8\,\Delta^3 - 2\,\frac{\phi(-\Delta)-\phi(0)}{\Delta^2} - 3\,\frac{\dot{\phi}(-\Delta)}{\Delta} - \\ \qquad -2\,\phi^{(2)}(-\Delta) = \frac{\partial f(t_0,\phi(0),\psi_\Delta(\cdot))}{\partial t} + \\ \qquad + \frac{\partial f(t_0,\phi(0),\psi_\Delta(\cdot))}{\partial x}\, f(t_0, \phi(0), \psi_\Delta(\cdot)) + \\ \qquad + \partial_y f(t_0, \phi(0), \psi_\Delta(\cdot))\,. \end{cases} \tag{16.13}$$

Note, right parts of equation (16.11) also depend on β_1 and β_2.

16.4 Sewing procedure of the second order for linear delay differential equation

Consider the sewing procedure for the linear equation with the distributed delay

$$\dot{x}(t) = a\,x(t) + b\,x(t-\tau) + \int_{-\tau}^{0} c(s)\,x(t+s)\,ds\,. \quad (16.14)$$

The right part of the equation (16.14) is the functional

$$f(x, y(\cdot)) = a\,x + b\,y(-\tau) + \int_{-\tau}^{0} c(s)\,y(s)\,ds\,,$$

hence for the function (16.10) one can calculate

$$f(\phi(0), \psi_\Delta(\cdot)) = a\,\phi(0) + b\,\phi(-\tau) +$$

$$+ \int_{-\tau}^{-\Delta} c(s)\,\phi(s)\,ds + \int_{-\Delta}^{0} c(s)\,\psi(s)\,ds =$$

$$= a\,\phi(0) + b\,\phi(-\tau) + \int_{-\tau}^{-\Delta} c(s)\,\phi(s)\,ds +$$

$$+ \int_{-\Delta}^{0} c(s)\left[\phi(0) + \alpha_1 s + \alpha_2\,s^2 + \alpha_3\,s^3 + \beta_1\,s^4 + \beta_2\,s^5\right] ds =$$

$$= a\,\phi(0) + b\,\phi(-\tau) + \int_{-\tau}^{-\Delta} c(s)\,\phi(s)\,ds + \phi(0)\int_{-\Delta}^{0} c(s)\,ds +$$

$$+\,\alpha_1 \int\limits_{-\Delta}^{0} c(s)\, s\, ds \,+\alpha_2 \int\limits_{-\Delta}^{0} c(s)\, s^2\, ds \,+\alpha_3 \int\limits_{-\Delta}^{0} c(s)\, s^3\, ds +$$

$$+\,\beta_1 \int\limits_{-\Delta}^{0} c(s)\, s^4\, ds \,+\beta_2 \int\limits_{-\Delta}^{0} c(s)\, s^5\, ds \;=$$

$$=\alpha_1\, A_1 + \alpha_2\, A_2 + \alpha_3\, A_3 + \beta_1\, B_1 + \beta_2\, B_2 +$$

$$+\,\phi(0)\left[a + \int\limits_{-\Delta}^{0} c(s)\, ds\right] + b\,\phi(-\tau) + \int\limits_{-\tau}^{-\Delta} c(s)\,\phi(s)\, ds\,,$$

where

$$A_1 = \int\limits_{-\Delta}^{0} c(s)\, s\, ds\,,\quad A_2 = \int\limits_{-\Delta}^{0} c(s)\, s^2\, ds\,,\quad A_3 = \int\limits_{-\Delta}^{0} c(s)\, s^3\, ds\,,$$

$$B_1 = \int\limits_{-\Delta}^{0} c(s)\, s^4\, ds\,,\quad B_2 = \int\limits_{-\Delta}^{0} c(s)\, s^5\, ds\,.$$

One can calculate, that

$$\frac{\partial f(\phi(0), \psi_\Delta(\cdot))}{\partial x} = a\,,$$

$$\partial_y f(\phi(0), \psi_\Delta(\cdot)) =$$

$$= b\,\dot{\psi}_\Delta(-\tau) \,+\, c(0)\,\psi_\Delta(0) \,-\, c(-\tau)\,\psi_\Delta(-\tau) \;=$$

$$=\; c(0)\,\psi(0) \,-\, c(-\tau)\,\phi(-\tau) \,+\, b\,\dot{\phi}(-\tau) =$$

$$=\; c(0)\,\phi(0) \,-\, c(-\tau)\,\phi(-\tau) \,+\, b\,\dot{\phi}(-\tau)\,.$$

Thus in the case of linear system (16.14) the equations (16.12) have the form

$$\begin{cases} \alpha_1 = \alpha_1 A_1 + \alpha_2 A_2 + \alpha_3 A_3 + \beta_1 B_1 + \beta_2 B_2 + \\ \quad + \phi(0) \left[a + \int\limits_{-\Delta}^{0} c(s)\, ds \right] + b\,\phi(-\tau) + \int\limits_{-\tau}^{-\Delta} c(s)\,\phi(s)\, ds\,, \\ \alpha_2 = \alpha_1\, a\, A_1 + \alpha_2\, a\, A_2 + \alpha_3\, a\, A_3 + \beta_1\, a\, B_1 + \beta_2\, a\, B_2 + \\ \quad + a\,\phi(0) \left[a + \int\limits_{-\Delta}^{0} c(s)\, ds \right] + a\, b\,\phi(-\tau) + \\ \quad + a \int\limits_{-\tau}^{-\Delta} c(s)\,\phi(s)\, ds + c(0)\,\phi(0) - c(-\tau)\,\phi(-\tau) + b\,\phi'(-\tau)\,. \end{cases} \tag{16.15}$$

Therefore the values α_1, α_2, α_3, β_1, β_2 can be calculated from the system of linear algebraic equations (16.11), (16.15).

The determinant of this system is

$$\begin{pmatrix} 1 & 0 & 0 & \Delta^3 & 3\,\Delta^4 \\ 0 & 1 & 0 & -3\,\Delta^2 & -8\,\Delta^3 \\ 0 & 0 & 1 & -3\,\Delta & 6\,\Delta^2 \\ 1 - A_1 & -A_2 & -A_3 & -B_1 & -B_2 \\ a\,A_1 & 1 - a\,A_2 & -a\,A_3 & -a\,B_1 & -a\,B_2 \end{pmatrix}.$$

Chapter 2

Numerical methods for functional differential equations

In this chapter basing on the i-smooth analysis, we develop a general approach to constructing numerical methods for systems with delays.

We consider the system of functional differential equations

$$\dot{x}(t) = f(t, x(t), x_t(\cdot)), \tag{0.1}$$

($x_t(\cdot) = \{x(t+s), -\tau \leq s < 0\}$) with the initial conditions

$$x(t_0) = x_0, \tag{0.2}$$

$$x_{t_0}(\cdot) = \{y^0(s), -\tau \leq s < 0\}. \tag{0.3}$$

Here $f : [t_0, t_0 + \theta] \times \mathbb{R}^n \times Q[-\tau, 0) \to \mathbb{R}^n$; $\theta > 0$ is the length of the integrating interval.

For a fixed t the vector $x(t)$ is called the *finite dimensional component of the phase state*, the function prehistory $x_t(\cdot)$ is called an *infinite dimensional* . Once more we emphasize that in the framework of our approach the finite dimensional component is separated from the infinite dimensional component of the phase state [1].

Further we assume fulfillment of the following assumptions:

Condition 0.1 *The mapping f is continuous on shift (see definition 11.3).*

Condition 0.2 *The mapping f satisfies the Lipschitz condition with respect to x and y, i.e. there exist constants L, M such that for all $t \in [t_0, t_0 + \theta]$; $x^{(1)}, x^{(2)} \in \mathbb{R}^n$; $y^{(1)}(\cdot), y^{(2)}(\cdot) \in Q[-\tau, 0)$ the following inequality is valid*

$$\|f(t, x^{(1)}, y^{(1)}(\cdot)) - f(t, x^{(2)}, y^{(2)}(\cdot))\| \leq$$

$$\leq L\|x^{(1)} - x^{(2)}\| + M\|y^{(1)}(\cdot) - y^{(2)}(\cdot)\|_Q.$$

From the theorem 14.1 it follows that there exists $\hat{\Delta} > 0$ such that there exists the unique solution of initial value problem (0.1) - (0.3) on the interval $[t_0, t_0 + \hat{\Delta}]$.

The condition 0.2 can be relaxed to *local Lipschitz condition.*

Below we describe some numerical methods (numerical models) of solving FDE.

The general approach to numerical FDE methods basing on i-smooth analysis was developed by the author around 1995 in the following form:

1) *Stage 1*: Advancement of the finite dimensional component according to an ODE numerical scheme.

2) *Stage 2*: Handling (treatment) of the infinite dimensional component for preparation for the next advancement of the finite dimensional component.

First we illustrate features of our approach on a simplest numerical model — *the Euler method.*

---

[1] this presentation differs this FDE form from other usually considered FDE presentations

17 Numerical Euler method

Let us set on the interval $[t_0, t_0 + \theta]$ a grid $t_l = t_0 + l\Delta$, $l = 0, 1, ..., N$ with the uniform step $\Delta = \theta/N$, where N is a natural. For the sake of simplicity we assume that $\tau/\Delta = m$ is a natural. Introduce the *discrete numerical model* of the system (0.1) denoting by $u_l \in \mathbb{R}^n$ the approximation of the exact solution $x(t_l) = x_l$ at the point t_l.

In contradiction to ODE the right-hand side of the system (0.1) is the mapping f defined on functions (prehistories), therefore it is necessary to have a special procedure of calculating the mapping f using the finite number of points — the calculated values of the numerical model (the discrete approximate solution). Generally speaking learning the discrete prehistory is not sufficient for constructing a numerical model adequate for the system (0.1).

To calculate the functional f on an approximate solution[2] one can use interpolation[3].

The simplest method of interpolation is the piece-wise constant interpolation

$$u(t) = \begin{cases} u_i, \; t \in [t_i, t_{i+1}), \; i = 0, 1, ..., l, \\ y^0(t - t_0), \; t \in [t_0 - \tau, t_0). \end{cases} \tag{17.1}$$

Note that at moment t_l function-prehistory $u_{t_l}(\cdot) \in$ $\in Q[-\tau, 0)$ is defined if discrete model prehistory u_i, $i \leq l$ is given.

The *Euler method with piece-wise constant interpolation of the discrete prehistory* is the following step-by-step model

$$u_0 = x_0; \tag{17.2}$$

$$u_{l+1} = u_l + \Delta f(t_l, u_l, u_{t_l}(\cdot)), \; l = 0, ..., N-1, \tag{17.3}$$

where $u_{t_l}(\cdot) = \{u(t_l + s), -\tau \leq s < 0\}$ is the model prehistory, defined by the interpolation (17.1).

---

[2]finite set of points

[3]obtaining a function by a finite set of points

Let $x(t), t \in [t_0-\tau, \theta]$ be the exact solution of the system (0.1). We say that the numerical method *converges* if

$$\max_{1\le l\le N} \|u_l - x(t_l)\| \to 0 \text{ as } N \to \infty,$$

and has the *convergence order* p if there exists a constant C such that $\|u_l - x(t_l)\| \le C\Delta^p$ for all $l = 1, ..., N$ and constant C does not depend on N.

Our next goal is to show that the Euler method (17.1) - (17.3) converges and has the first order of convergence.

Lemma 17.1 *Let the exact solution $x(t)$ be a continuous differentiable function, then the piece-wise constant interpolation has the property*

$$\|x_{t_l}(\cdot) - u_{t_l}(\cdot)\|_Q \le \max_{l-m\le i\le l} \|x_i - u_i\| + C_1\Delta, \ \ l = 0, ..., N. \tag{17.4}$$

Proof. Let $\hat{x}(t)$ be the piece-wise constant function defined by the formula

$$\hat{x}(t) = (x(t_i), \ \ t \in [t_i, t_{i+1})).$$

Then

$$\|x_{t_l}(\cdot) - u_{t_l}(\cdot)\|_Q = \sup_{t_l-\tau\le t\le t_l} \|x(t) - u(t)\| \le$$

$$\le \sup_{t_l-\tau\le t\le t_l} \|x(t) - \hat{x}(t)\| + \sup_{t_l-\tau\le t\le t_l} \|\hat{x}(t) - u(t)\| \le$$

$$\le \max_{l-m\le i\le l} \|x_i - u_i\| + C_1\Delta,$$

where C_1 is the maximum of the absolute value of the derivative of the function $x(t)$ on the interval $[t_0, t_0+\theta]$. □

Definition 17.1 *Residual is the function*

$$\psi(t_l) = \frac{x_{l+1} - x_l}{\Delta} - f(t_l, x_l, x_{t_l}(\cdot)).$$

Lemma 17.2 *Let the exact solution $x(t)$ be twice the continuous differentiable function, then the residual of the Euler method has the first order with respect to Δ.*

Proof. Let us expend $x(t_{l+1})$ by the Taylor formular in a neighbourhood of the point t_l:

$$x(t_{l+1}) = x(t_l) + \Delta f(t_l, x_l, x_{t_l}(\cdot)) + \ddot{x}(c)\frac{\Delta^2}{2}, \quad c \in [t_l, t_{l+1}],$$

then for every $l = 0, \ldots, N-1$ the following condition is valid

$$\|\psi(t_l)\| \leq C_2\Delta,$$

where C_2 is the half of the maximum of the absolute value of the second derivative of the function $x(t)$ on the interval $[t_0, t_0 + \theta]$. □

Theorem 17.1 *If the conditions of lemma 17.2 and the conditions 0.1, 0.2 are fulfilled then the Euler method with piece-wise constant interpolation converges and has the first convergence order.*

Proof. Denote $\epsilon_l = \|x_l - u_l\|$ and express ϵ_{l+1} by ϵ_l, for this operation the value x_{l+1} substitute from the definition of residual and u_{l+1} from the step-by-step formula (17.3):

$$\epsilon_{l+1} = \|x_{l+1} - u_{l+1}\| =$$

$$= \|x_l + \Delta f(t_l, x_l, x_{t_l}(\cdot)) + \Delta\psi(t_l) - u_l - \Delta f(t_l, u_l, u_{t_l}(\cdot))\| \leq$$

$$\leq \epsilon_l + \Delta\|f(t_l, x_l, x_{t_l}(\cdot)) - f(t_l, u_l, u_{t_l}(\cdot))\| + \Delta\|\psi(t_l)\|.$$

Utilizing condition 0.2 (Lipshitz condition for f) and Lemma 17.2, we obtain the estimate

$$\epsilon_{l+1} \leq \epsilon_l + \Delta(L\|x_l - u_l\| + M\|x_{t_l}(\cdot) - u_{t_l}(\cdot)\|_Q) + C_2\Delta^2,$$

which together with Lemma 17.1 give

$$\epsilon_{l+1} \leq \epsilon_l(1 + L\Delta) + \Delta M \max_{l-m \leq i \leq l} \|\epsilon_i\| + \Delta^2(MC_1 + C_2). \tag{17.5}$$

By mathematical induction method l

$$\epsilon_l \leq (1 + \Delta(L + M + 1))^l (MC_1 + C_2)\Delta. \tag{17.6}$$

Basis: $\epsilon_0 = 0$.

Induction Step.

Let the max in the right side of estimate (17.5) be achieved at an index $i_0 \leq l$ and then by applying the inductive assumption to ϵ_l and ϵ_{i_0} we obtain

$$\epsilon_{l+1} \leq \epsilon_l(1 + L\Delta) + \Delta M \epsilon_{i_0} + \Delta^2(MC_1 + C_2) \leq$$
$$\leq (1 + \Delta(L + M + 1))^l (MC_1 + C_2)\Delta(1 + L\Delta) +$$
$$+\Delta M(1+\Delta(L+M+1))^{i_0}(MC_1+C_2)\Delta+\Delta^2(MC_1+C_2) \leq$$
$$\leq (1 + \Delta(L + M + 1))^l (MC_1 + C_2)\Delta(1 + L\Delta + M\Delta + \Delta).$$

The estimation (17.6) is proven.

Because $N = \frac{\theta}{\Delta}$, then for all $l = 0, \ldots, N$ we have the estimation

$$\epsilon_l \leq (1 + \Delta(L + M + 1))^{\frac{\theta}{\Delta}} (MC_1 + C_2)\Delta \leq$$
$$\leq e^{(L+M+1)\theta}(MC_1 + C_2)\Delta.$$

This estimate contains the claim of the theorem. □

The described Euler method with the piece-wise constant interpolation is the simplest among convergence methods. To obtain a better convergence one can use a more complicated interpolation and get a more exact approximation by complification of step-by-step model.

18 Numerical Runge-Kutta-like methods

The Runge-Kutta numerical methods are the most effective and popular computational procedures of solving wide classes of ordinary differential equations. Consider the application of i-Smooth analysis to elaborating the Runge-Kutta-like methods for functional differential equations of the general form.

18.1 Methods of interpolation and extrapolation

This section is devoted to the general discussion of the interpolation procedure. Let us consider something similar to the previous section, a defined grid on the interval $[t_0, t_0 + \theta]$. *Discrete prehistory of the model at a moment* t_l is the set of $m + 1$ vectors:

$$\{u_i\}_l = \{u_i \in \mathbb{R}^n, \quad l - m \le i \le l\}.$$

This set at the moment t_l defines future behaviour of the discrete model.

Definition 18.1 *An operator of interpolation I of the discrete prehistory of the model is a mapping $I : \{u_i\}_l \to u(\cdot) \in Q[t_l - \tau, t_l]$.*

Definition 18.2 *Operator I has the approximation order p at the exact solution if there exist constants C_1, C_2 such that for all $l = 0, 1, \ldots, N$ and $t \in [t_l - \tau, t_l]$ the following inequality is hold*

$$\|x(t) - u(t)\| \le C_1 \max_{l-m \le i \le l} \|u_i - x_i\| + C_2 \Delta^p. \tag{18.1}$$

The interpolation by piece-wise linear curves:

$$I : \{u_i\}_l \to u(t) =$$

$$= \begin{cases} ((t - t_i)u_{i+1} + (t_{i+1} - t)u_i)\frac{1}{\Delta}, t \in [t_i, t_{i+1}], \ i = 0, \ldots, l-1, \\ y^0(t_0 - t), \ t \in [t_l - \tau, t_0). \end{cases} \tag{18.2}$$

is an example of the operator of the second approximation order on sufficiently smooth solutions.

Consider a more general method of interpolating by functions, piece-wise composed by polynomials of a degree $p \in \mathbb{N}$. These functions are called *degenerate splines*. Without loss of generality one can assume that $\frac{m}{p} = k$ is a natural, otherwise one can take m multiple (divisible) to p.

Let $t_l \in [t_0, t_0 + \theta]$, $t_i = t_0 + i\Delta$, $i = l - m, \ldots, l$ are points of the grid, and $\{ u_i,\ i = l - m, \ldots, l \}$ is the corresponding discrete prehistory of the model. If $t_i < t_0$, then $u_i = y^0(t_0 - t_i)$; for $t_i = t_0$: $u_0 = x_0$.

Let us divide the interval $[t_l - \tau, t_l]$ on k subintervals $[t_{l_{i+1}}, t_{l_i}]$, $i = 0, 1, \ldots, k - 1$ of the length $p\Delta$ such that $t_{l_0} = t_l$, $t_{l_1} = t_{l-p}$, ... On every interval $[t_{l_{i+1}}, t_{l_i}]$ we construct the interpolation polynomial $L_p(t) = L_p^i(t)$ by points $u_{l_i-p}, u_{l_i-p+1}, \ldots, u_{l_i}$:

$$L_p^i(t) = \sum_{n=0}^{p} u_{l_i-n} \prod_{j=l_i-p;\ j\neq l_i-n}^{l_i} \frac{t - t_j}{t_{l_i-n} - t_j}.$$

Definition 18.3 *The interpolation operator of the model prehistory by degenerate splines of the degree p is the mapping*

$$I : \{u_i\}_l \to u(t) = \begin{cases} L_p^i(t), \ t_{l_{i+1}} \leq t < t_{l_i},\ i = 0, \ldots, l - 1, \\ y^0(t_0 - t),\ t \in [t_l - \tau, t_0). \end{cases} \tag{18.3}$$

Theorem 18.1 *Let the exact solution $x(t)$ be $(p+1)$-time continuous differentiable on the interval $[t_0 - \tau, t_0 + \theta]$ then the interpolation operator by degenerate spline of p-th degree has the approximation order of order $p + 1$.*

Proof. Fix the arbitrary $l = 0, \ldots, N$ and $t \in [t_l - \tau, t_l]$. Let t belongs to a subinterval $[t_{l_{i+1}}, t_{l_i}]$. Then

$$\|x(t) - u(t)\| = \|x(t) - L_p^i(t)\| \leq$$

$$\leq \|x(t) - \hat{L}_p^i(t)\| + \|L_p^i(t) - \hat{L}_p^i(t)\|,$$

where $\hat{L}_p^i(t)$ the interpolational polynomial of a degree p is constructed by values x_j on the interval $[t_{l_{i+1}}, t_{l_i}]$:

$$\hat{L}_p^i(t) = \sum_{n=0}^{p} x_{l_i-n} \prod_{j=l_{i+1};j\neq l_i-n}^{l_i} \frac{t - t_j}{t_{l_i-n} - t_j}.$$

The approximation error of the exact solution can be calculated by the formula [31, p. 133]

$$\|R_p(t)\| = \|x(t) - \hat{L}_p^i(t)\| = \|\frac{x^{(p+1)}(\xi)}{(p+1)!} \prod_{j=0}^{p} (t - t_{l_i - j})\|,$$

where $\xi \in [t_{l_{i+1}}, t_{l_i}]$.

Thus, if $t \in [t_{l_{i+1}}, t_{l_i}]$, then

$$\| \prod_{j=0}^{p} (t - t_{l_i - j})\| \le \Delta \cdot \Delta \cdot (2\Delta) \cdots (p\Delta) = \Delta^{p+1} p!$$

and

$$\|R_p(t)\| \le \frac{M_{p+1}}{(p+1)!} p!\, \Delta^{p+1} = \frac{M_{p+1}}{p+1} \Delta^{p+1}, \tag{18.4}$$

where $M_{p+1} = \max\limits_{t_0 - \tau \le t \le t_0 + \theta} \|x^{(p+1)}(t)\|$.

Besides, we have the estimation

$$\|L_p^i(t) - \hat{L}_p^i(t)\| =$$

$$= \sum_{n=0}^{p} \|u_{l_i - n} - x_{l_i - n}\| \prod_{j = l_i - p; j \ne l_i - n}^{l_i} \frac{|\, t - t_j \,|}{|\, t_{l_i - n} - t_j \,|} \le$$

$$\le \max_{l_i - p \le n \le l_i} \|u_n - x_n\| \, (p+1)!. \tag{18.5}$$

From (18.4) and (18.5) obtain

$$\|x(t) - L_p^i(t)\| \le C_1 \max_{l_i - k \le n \le l_i} \|u_n - x_n\| + C_2 \Delta^{p+1}. \tag{18.6}$$

Here $C_1 = (p+1)!$, $C_2 = \dfrac{M_{p+1}}{p+1}$. □

Consider a modification of the interpolation by degenerate splines. This modification can be called the *interpolation by the nearest points.*

Let t_l be fixed, $1 \le p \le m$ is a natural. On every interval $[t_i, t_{i+1}]$, $i = l - m, \ldots, l - 1$, we construct a polynomial $L_p^i(t)$ by the following rule:

$(p+1)$-th point of the interpolation t_{i_j} is defined by the relations: $t_{i_0} = t_i$, $t_{i_1} = t_{i+1}$, $t_{i_2} = t_{i-1}$, $t_{i_3} = t_{i+2}$ and so on, if $t_{i_j} \leq t_l$. If applying this algorithm, we obtain for some j that $t_{i_j} > t_l$[4], then we take as t_{i_j} the nearest unused point (knot) to the left of the interval $[t_i, t_{i+1}]$.

For example, if $t_{i_3} = t_{i+2} > t_l$ (this will be when $i = l-1$), then we take t_{i-2} as t_{i_3}.

Denoting the model prehistory

$$\prod_{k=0;\ k\neq j}^{p} \frac{t - t_k}{t_{i_j} - t_k}.$$

If $t_{i_j} < t_0$, then $u_{i_j} = y^0(t_0 - t_{i-j})$. In case of $t_{i_j} = t_0$: $u_0 = x_0$.

Define the interpolation operator of the model prehistory

$$I : \{u_i\}_l \to u(t) = \begin{cases} L_p^i(t), \ t_i \leq t < t_{i+1}, \ i = 0, \ldots, l-1, \\ y^0(t_0 - t), \ t \in [t_l - \tau, t_0). \end{cases}$$

Theorem 18.1 is valid for this method of interpolation: moreover, in some cases the estimation in the proof of the theorem can be improved

Lemma 18.1 *Let conditions of theorem 18.1 be satisfied and $[t_i, t_{i+1}] \subset [t_l - \tau, t_l]$, moreover, $t_i + (p-1)\Delta \leq t_l$ for odd p or $t_i + (p-2)\Delta \leq t_l$ for even p. Then the interpolational operator by nearest points has the property:*

For any $t \in [t_i, t_{i+1}]$

$$\|x(t) - u(t)\| \leq \hat{C}_1 \max_{l-m \leq n \leq l} \|u_n - x_n\| + \hat{C}_2 \Delta^{p+1},$$

where for $p = 2k+1$

$$\hat{C}_1 = (2k+2)((k+1)!)^2, \ \hat{C}_2 = \frac{((k+1)!)^2 M_{p+1}}{(2k+2)!},$$

[4] i.e. overcoming the right point

for $p = 2k$

$$\hat{C}_1 = (2k+1)(k!(k+1)!, \ \hat{C}_2 = \frac{k!(k+1)M_{p+1}}{(2k+1)!}.$$

Proof. Fix arbitrary $t \in [t_i, t_{i+1}] \subset [t_l - \tau, t_l]$. It follows that for all knots of interpolation we have $t_{i_j} \le t_l$. Then

$$\|x(t) - u(t)\| = \|x(t) - L_p^i(t)\| \le$$

$$\le \|x(t) - \hat{L}_p^i(t)\| + \|L_p^i(t) - \hat{L}_p^i(t)\|,$$

where $\hat{L}_p^i(t)$ is the interpolational polynomial of the degree p, constructed according to the values $x_{i_j} = x(t_{i_j})$:

$$L_p^i(t) = \sum_{j=0}^{p} x_{i_j} \prod_{n=0;\ n\neq j}^{p} \frac{t - t_k}{t_{i_j} - t_n}.$$

The approximation error of the interpolation on the exact solution is calculated by the formula

$$\|R_p(t)\| = \|x(t) - \hat{L}_p^i(t)\| = \|\frac{x^{(p+1)}(\xi)}{(p+1)!} \prod_{j=0}^{p} (t - t_{i_j})\|,$$

where $\xi \in [t_{l-m}, t_l]$.

Consider two cases:

1) p is an even number, i.e. $p = 2k+1$.

If $t \in [t_i, t_{i+1}]$, then

$$\| \prod_{j=0}^{p} (t - t_{i_j})\| \le \Delta \cdot \Delta \cdot (2\Delta)(2\Delta) \cdots ((k+1)p\Delta)((k+1)p\Delta) =$$

$$= \Delta^{p+1}((k+1)!)^2$$

and

$$\|R_p(t)\| \le \frac{M_{p+1}((k+1)!)^2}{2k+2} \Delta^{p+1},$$

where $M_{p+1} = \max\limits_{t_0-\tau \leq t \leq t_0+\theta} \|x^{(p+1)}(t)\|$. Besides this we have the estimate

$$\|L_p^i(t) - \hat{L}_p^i(t)\| = \sum_{j=0}^{p} \|u_{i_j} - x_{i_j}\| \prod_{n=0;n\neq i_j}^{p} \frac{| t - t_n |}{| t_{i_j} - t_n |} \leq$$

$$\leq \max_{l-m \leq n \leq l} \|u_n - x_n\| \, (p+1)((k+1)!)^2.$$

from which follows the statement of the lemma for even p.

2) p is an odd number, i.e. $p = 2k$. In this case be some changes in estimates

$$\|\prod_{j=0}^{p}(t-t_{i_j})\| \leq \Delta{\cdot}\Delta{\cdot}(2\Delta)(2\Delta){\cdots}(kp\Delta)(kp\Delta)((k+1)p\Delta) =$$

$$= \Delta^{p+1} k!(k+1)!$$

and

$$\prod_{n=0;n\neq i_j}^{p} \frac{| t - t_n |}{| t_{i_j} - t_n |} \leq k!((k+1)!.$$

This proofs the lemma for odd p. □

The corresponding estimates can be obtained when only a subset of interpolation knots satisfies the conditions $t_{i_j} \leq t_l$. These estimates give smaller constants than C_1 and C_2 in the general case of theorem 18.1.

Besides the interpolation by degenerate splines, in the described methods below one can use the *Hermite interpolation with multiple knotes* $\dot{u}_l = f(t_l, u_l, u_{t_l})$, and other methods of interpolation of the discrete model prehistory.

As is described below in numerical methods, it is necessary at a moment t_l to know the prehistory of the model $u_{t_l+a\Delta}(\cdot)$ for $a > 0$. This requires realizing an extrapolation of the model on the interval $[t_l, t_l + a\Delta]$.

Definition 18.4 *For a constant $a > 0$ an extrapolation operator E of the model prehistory is a mapping*

$$E : \{u_i\}_l \to u(\cdot) \in Q[t_l, t_l + a\Delta] \, .$$

Definition 18.5 *An extrapolation of the model prehistory has the p-th order of the approximation error on the exact solution, if there exist constants* C_3 *and* C_4 *such that for all* $a > 0$ *and* $l = 0, 1, \ldots, N-1$, $t \in [t_l, t_l + a\Delta]$ *the following inequality is satisfied*

$$\|x(t) - u(t)\| \leq C_3 \max_{l-m \leq i \leq l} \|u_i - x_i\| + C_4(\Delta)^p. \quad (18.7)$$

A method of defining an extrapolational operator is the extrapolation by *continuation* of the interpolational polynomial

$$E : \{u_i\}_l \to u(t) = L_p^0(t), \quad t \in [t_l, t_l + a\Delta], \quad (18.8)$$

where $L_p^0(t)$ is the interpolational polynomial of p-th degree, constructed by values u_j on the interval $[t_{l-p}, t_l]$:

$$L_p^0(t) = \sum_{n=0}^{p} u_{l-n} \prod_{j=l-p; j \neq l-n}^{l} \frac{t - t_j}{t_{l-n} - t_j}.$$

Theorem 18.2 *Let the exact solution* $x(t)$ *be* $(p+1)$*-time continuous differentiable on an interval* $[t_0 - \tau, t_0 + \theta]$, *then the extrapolation operator by continuation of polynomial p-th degree has the approximation error of the order* $p + 1$.

Proof. Fix arbitrary $t \in [t_l, t_l + a\Delta]$. Then

$$\|x(t) - u(t)\| = \|x(t) - L_p^0(t)\| \leq$$

$$\leq \|x(t) - \hat{L}_p^0(t)\| + \|L_p^0(t) - \hat{L}_p^0(t)\|,$$

where $\hat{L}_p^0(t)$ is the interpolational polynomial of p-th degree, constructed by values x_j on the interval $[t_{l-p}, t_l]$.

Then

$$\|x(t) - \hat{L}_p^o(t)\| \leq \frac{M_{p+1}}{(p+1)!} \prod_{j=0}^{p} (t - t_{l-j}) \leq$$

$$\leq \frac{M_{p+1}}{(p+1)!} \prod_{j=0}^{p} (a\Delta + j\Delta) \leq C_4(\Delta)^{p+1},$$

where

$$C_4 = \frac{M_{p+1}}{(p+1)!} \prod_{j=0}^{p} (a+j).$$

Besides

$$\|L_p^0(t) - \hat{L}_p^0(t)\| \leq$$

$$\leq \max_{l-p \leq n \leq l} \|u_n - x_n\| \sum_{n=0}^{p} \prod_{j=0; j \neq n}^{p} \frac{(a\Delta + j\Delta)}{\Delta} \leq$$

$$\leq C_3 \max_{l-p \leq n \leq l} \|u_n - x_n\|,$$

where

$$C_3 = (p+1) \prod_{j=0}^{p} (a+j).$$

The theorem is proved. □

Another extrapolation method is based on the application of continuous numerical methods [5].

For example, extrapolation by the Euler method

$$E : \{u_i\}_l \to u(t) = u_l + (t - t_l) f(t_l, u_l, u_{t_l}), t \in [t_l, t_l + a\Delta]$$

has the approximation order $p = 2$ under the condition that $x(\cdot)$ is sufficiently smooth.

In some cases it is convenient to unify operators of interpolation and extrapolation into the one interpolation-extrapolation operator.

Definition 18.6 *For some $a > 0$ the interpolation-extrapolation operator IE of the discrete model prehistory is a mapping*

$$IE : \{u_i\}_l \to u(\cdot) \in Q[t_l - \tau, t_l + a\Delta].$$

[5]continuous numerical methods for FDE we discuss further in the book. In this case the approximation order p is equal to a local approximation error of the continuous method.

Definition 18.7 *IE operator has an approximation order p on the exact solution, if constants exist C_5, C_6 such that for all $a > 0$, $l = 0, 1, \ldots, N-1$ and $t \in [t_l - \tau, t_l + a\Delta]$ the following inequality is valid*

$$\|x(t) - u(t)\| \leq C_5 \max_{l-m \leq i \leq l} \|u_i - x_i\| + C_6(\Delta)^p. \quad (18.9)$$

18.2 Explicit Runge-Kutta-like methods

Below we assume that operators a interpolation operator I and an extrapolation operator E are fixed.

Definition 18.8 *For a natural k the k-stage explicit Runge-Kutta-like method – ERK (with interpolation I and extrapolation E) is the numerical model*

$$u_0 = x_0, \quad (18.10)$$

$$u_{l+1} = u_l + \Delta \sum_{i=1}^{k} \sigma_i \, h_i(u_l, u_{t_l}(\cdot)), \; l = 0, \ldots, N-1, \quad (18.11)$$

$$h_1(u_l, u_{t_l}(\cdot)) = f(t_l, u_l, u_{t_l}(\cdot)), \quad (18.12)$$

$$h_i(u_l, u_{t_l}(\cdot)) = f(t_l + a_i\Delta, u_l + \Delta \sum_{j=1}^{i-1} b_{ij} h_j(u_l, u_{t_l}(\cdot)), u_{t_l + a_i\Delta}(\cdot)). \quad (18.13)$$

Here the prehistory of the model is defined by relations

$$u_t(s) = \begin{cases} y^0(t + s - t_0) & \text{for } t + s < t_0, \\ I(\{u_i\}_l) & \text{for } t_l - \tau \leq t + s < t_l, \\ E(\{u_i\}_l) & \text{for } t_l \leq t + s \leq t_l + a\Delta, \end{cases} \quad (18.14)$$

$$a = \max\{|a_i|, 1 \leq i \leq k\}.$$

Numbers a_i, σ_i, b_{ij} are called *coefficients* of the method. We denote by $\sigma = \max\{|\sigma_i|\}, b = \max\{|b_{ij}|\}$.

For ODE the coefficients of the method are usually exposed in the *Butcher table.* In the framework of our approach the coefficients of the method are defined similar to ODE case.

$$
\begin{array}{c|cccc}
0 & 0 & 0 & \dots & 0 \\
a_2 & b_{21} & 0 & \dots & 0 \\
. & . & . & \dots & . \\
a_k & b_{k1} & b_{k2} & \dots & 0 \\
\hline
\end{array}
$$

Our approach also allows one similar to ODE investigate the convergence order (see definitions of section 22) of the ERK-method.

Definition 18.9 *The residual (approximation error) of the ERK-method is the function*

$$\psi(t_l) = \frac{x_{l+1} - x_l}{\Delta} - \sum_{i=1}^{k} \sigma_i \, h_i(x_l, x_{t_l}(\cdot)).$$

Note, the residual is defined on the exact solution $x(t)$ and does not depend on interpolation and extrapolation.

Definition 18.10 *Residual has the order p if there exists a constant C such that $\|\psi(t_l)\| \leq C\Delta^p$ for all $l = 0, 1, ..., N-1$.*

Theorem 18.3 *If the method (18.10) - (18.14) has the residual of the order $p_1 > 0$, the interpolation of the model prehistory has the order $p_2 > 0$, the extrapolation of the model prehistory has the order $p_3 > 0$, then the method converges. Moreover, the convergence order p of the ERK-method is not less than minimum of p_1, p_2, p_3.*

To prove the theorem we use two auxiliary statements

Lemma 18.2 *Functionals h_i defined by (18.12) - (18.13), satisfy the following Lipschitz condition:*

There exist constants L_i and M_i, such that

$$\|h_i(u_l, u_{t_l}(\cdot)) - h_i(x_l, x_{t_l}(\cdot))\| \leq$$

$$\leq L_i\|u_l - x_l\| + M_i \sup_{t_l-\tau\leq t\leq t_l+a\Delta} \|u(t) - x(t)\|.$$

Proof We prove the lemma by the induction method with respect to i.

For $i = 1$ we obtain the functional $h_1(u_l, u_{t_l}(\cdot)) = f(t_l, u_l, u_{t_l}(\cdot))$ which according to condition 0.2 satisfies the Lipschitz condition with $L_1 = L, M_1 = M$.

Assume that for indices $j \leq i-1$, functionals h_j satisfy the Lipschitz condition with constants L_i, M_i. Show that the functionals also satisfy the Lipschitz condition:

$$\|h_i(u_l, u_{t_l}(\cdot)) - h_i(x_l, x_{t_l}(\cdot))\| =$$

$$= \|f(t_l + a_i\Delta, u_l + \Delta\sum_{j=1}^{i-1} b_{ij}h_j(u_l, u_{t_l}(\cdot)), u_{t_l+a_i\Delta}(\cdot))-$$

$$-f(t_l + a_i\Delta, x_l + \Delta\sum_{j=1}^{i-1} b_{ij}h_j(x_l, x_{t_l}(\cdot)), x_{t_l+a_i\Delta}(\cdot))\| \leq$$

$$\leq L\|u_l - x_l\| + L\Delta b\sum_{j=1}^{i-1} \|h_j(u_l, u_{t_l}(\cdot)) - h_j(x_l, x_{t_l}(\cdot))\|+$$

$$+M \sup_{t_l-\tau\leq t\leq t_l+a\Delta} \|u(t) - x(t)\| \leq$$

$$\leq L\|u_l - x_l\| + L\Delta b\sum_{j=1}^{i-1}(L_j\|u_l - x_l\|+$$

$$+M_j \sup_{t_l-\tau\leq t\leq t_l+a\Delta} \|u(t)-x(t)\|)+M \sup_{t_l-\tau\leq t\leq t_l+a\Delta} \|u(t)-x(t)\|.$$

Thus the functional h_i satisfies the Lipschitz condition with constants

$$L_i = L + L\Delta b\sum_{j=1}^{i-1} L_j, \quad M_i = M + L\Delta b\sum_{j=1}^{i-1} M_j.$$

□

Lemma 18.3 *Let the interpolational operator have the approximation order p_2 and the extrapolational operator have the order p_3, then there exist constants $\hat{C}_1$ $\hat{C}_2$ such that the following inequalities are satisfied*

$$\|h_i(u_l, u_{t_l}(\cdot)) - h_i(x_l, x_{t_l}(\cdot))\| \leq$$

$$\leq \hat{C}_1 \max_{l-m \leq i \leq l} \|u_i - x_i\| + \hat{C}_2 \Delta^{\min\{p_2, p_3\}}.$$

Proof. Note, because the interpolational operator has the approximation order p_2 and the extrapolational operator has the order p_3, hence the IE-operator has the order $p = \min\{p_2, p_3\}$. Then from (18.9) and previous lemma it follows that

$$\|h_i(u_l, u_{t_l}(\cdot)) - h_i(x_l, x_{t_l}(\cdot))\| \leq$$

$$\leq L_i \|u_l - x_l\| + M_i \sup_{t_l - \tau \leq t \leq t_l + a\Delta} \|u(t) - x(t)\| \leq$$

$$\leq L_i \|u_l - x_l\| + M_i (C_5 \max_{l-m \leq i \leq l} \|u_i - x_i\| + \hat{C}_6 \Delta^{\min\{p_2, p_3\}}).$$

Taking

$$\hat{C}_1 = \max_{1 \leq i \leq k} M_i C_5 + \max_{1 \leq i \leq k} L_i, \quad \hat{C}_2 = \max_{1 \leq i \leq k} M_i C_6,$$

we obtain the validity of the lemma. □

Proof of theorem 18.3. Denote $\epsilon_l = \|x_l - u_l\|$ and express ϵ_{l+1} throught ϵ_l substituting x_{l+1} from definition of the residual, and u_{l+1} from step-by-step formula (18.11). Taking into account the Lemma 18.3 we obtain:

$$\epsilon_{l+1} = \|x_{l+1} - u_{l+1}\| =$$

$$= \|x_l + \Delta\psi(t_l) - u_l + \Delta \sum_{i=1}^{k} \sigma_i \left(h_i(x_l, x_{t_l}(\cdot)) - h_i(u_l, u_{t_l}(\cdot))\right)\| \leq$$

$$\leq \epsilon_l + \Delta\|\psi(t_l)\| + \Delta \sum_{i=1}^{k} |\sigma_i| \left(\hat{C}_1 \max_{l-m \leq i \leq l} \epsilon_i + \hat{C}_2 \Delta^{\min\{p_2, p_3\}}\right) \leq$$

$$\leq \epsilon_l + \Delta k\sigma\hat{C}_1 \max_{l-m\leq i\leq l} \epsilon_i + k\sigma\hat{C}_2\Delta^{\min\{p_2,p_3\}+1} + C\Delta^{p_1+1}.$$

Thus we have the estimation

$$\epsilon_{l+1} \leq \epsilon_l + \Delta C_7 \max_{l-m\leq i\leq l} \epsilon_i + C_8\Delta^{p+1}, \tag{18.15}$$

where $C_7 = k\sigma\hat{C}_1$, $C_8 = k\sigma\hat{C}_2 + C$, $p = \min\{p_1, p_2, p_3\}$.

Now let us apply the induction method with respect to l to prove the estimate

$$\epsilon_l \leq (1 + \Delta(C_7 + 1))^l C_8\Delta^p. \tag{18.16}$$

The basis of the induction. For $l = 0$ the estimate is valid, because $\epsilon_0 = 0$.

The induction step. Let the estimation (18.16) be valid for all indicis $\leq l$ [6].

Let the max on the right side of the estimation (18.15) be achievable at an index $i_0 \leq l$. Then apply the inductive assumption to ϵ_l and ϵ_{i_0} to obtain

$$\epsilon_{l+1} \leq \epsilon_l + \Delta C_7\epsilon_{i_0} + C_8\Delta^p \leq$$

$$\leq (1 + \Delta(C_7 + 1))^l C_8\Delta^p +$$

$$+\Delta C_7(1 + \Delta(C_7 + 1))^{i_0} C_8\Delta^p + C_8\Delta^p \leq$$

$$\leq (1 + \Delta(C_7 + 1))^l C_8\Delta^p(1 + C_7\Delta + \Delta).$$

The estimation (18.16) is proven.

From this estimation one can deduce (similar to theorem 17.1) the estimation

$$\epsilon_l \leq e^{(C_7+1)\theta} C_8\Delta^p,$$

which contains the claim of the theorem. □

[6] and we have to show it for $l + 1$

18.3 Order of the residual of ERK-methods

For ordinary differential equations the residual order of ERK-methods is defined by the expansion of the exact solution into the Taylor series.

Example 18.1 Consider for the ODE system

$$\dot{x} = f(t, x)$$

the Hoin method, [7]:

$$u_{l+1} = u_l + \frac{\Delta}{2}(f(t_l, u_l) + f(t_l, u_l + \Delta f(t_l, u_l))).$$

Residual of this method is defined by the relation

$$\psi(t_l) = \frac{x_{l+1} - x_l}{\Delta} - \frac{1}{2}(f(t_l, x_l) + f(t_l, x_l + \Delta f(t_l, x_l))).$$

Let us prove that the residual of the Hoin method has the second order on sufficiently smooth solutions.

Expanding $x(t)$ by the Taylor formula we have

$$x_{l+1} = x_l + \dot{x}(t_l)\Delta + \ddot{x}(t_l)\frac{\Delta^2}{2} + O(\Delta^3) =$$

$$= x_l + f(t_l, x_l)\Delta + \frac{\Delta^2}{2}\left[\frac{\partial f}{\partial t} + \frac{\partial f}{\partial x} f\right] + O(\Delta^3),$$

where $O(\Delta^3)$ is the value of the third order of smallness with respect to Δ and,

$$\frac{\partial f}{\partial x} = \begin{pmatrix} \frac{\partial f_1}{\partial x_1} & \frac{\partial f_1}{\partial x_2} & \cdots & \frac{\partial f_1}{\partial x_l} \\ \frac{\partial f_2}{\partial x_1} & \frac{\partial f_2}{\partial x_2} & \cdots & \frac{\partial f_2}{\partial x_l} \\ \vdots & \vdots & \ddots & \vdots \\ \frac{\partial f_n}{\partial x_1} & \frac{\partial f_n}{\partial x_2} & \cdots & \frac{\partial f_n}{\partial x_l} \end{pmatrix}$$

is the Jacobi matrix at the point (t_l, x_l).

[7] also called the improved Euler method

Besides, the following presentation follows from the Taylor formula

$$f(t_l, x_l + \Delta f(t_l, x_l)) = f(t_l, x_l) + \left[\frac{\partial f}{\partial t} + \frac{\partial f}{\partial x} f\right] \Delta + O(\Delta^2).$$

Substituting this expansion into the residual, we obtain $\psi(t_l) = O(\Delta^2)$.

The methodology of estimating the order of the residual for functional differential equations is also based on the expansion of a solution into the Taylor series, however in this case, the problem arises of calculating derivatives of vector functional (0.1) $f(t, x, y(\cdot))$ on the right side of the system (0.1). To overcome the problem we use the technique of i-Smooth calculus.

To obtain the Taylor series coefficients we use the coinvariant derivatives $\partial_t f(t_l, x(t_l), x_{t_l}(\cdot))$.

As the example, consider for the equation (0.1) a formula which is the analog of the improved Euler method

$$u_0 = x^0,$$

$$u_{l+1} = u_l + \frac{\Delta}{2}\Big(f(t_l, u_l, u_{t_l}(\cdot)) +$$

$$+ f(t_l + \Delta, u_l + \Delta f(t_l, u_l, u_{t_l}(\cdot)), u_{t_k+\Delta}(\cdot))\Big),$$

where $u_t(\cdot)$ is the result of acting on some operator of interpolation and extrapolation.

Describe the procedure of estimating the approximation order. Consider the residual of the method

$$\psi(t_l) = \frac{x(t_{l+1}) - x(t_l)}{\Delta} - \frac{1}{2}\Big(f(t_l, x(t_l), x_{t_l}(\cdot)) +$$

$$+ f(t_l, x(t_l) + \Delta f(t_l, x(t_l), x_{t_l}), x_{t_{l+1}})\Big).$$

Expanding the exact solution into the Taylor series we obtain

$$x(t_{l+1}) = x(t_l) + \dot{x}(t_l)\Delta + \ddot{x}(t_l)\frac{\Delta^2}{2} + O(\Delta^3) =$$

$$= x(t_k) + f(t_l, x(t_l), x_{t_l}(\cdot))\Delta +$$

$$+ \frac{\Delta^2}{2}\left[\partial_t f(t_l, x(t_l), x_{t_l}(\cdot)) + \frac{\partial f(t_l, x(t_l), x_{t_l}(\cdot))}{\partial x} f(t_l, x(t_l), x_{t_l}(\cdot))\right] +$$

$$+O(\Delta^3)\,.$$

Also we have

$$f(t_l + \Delta, x(t_l) + \Delta f(t_l, x(t_l), x_{t_l}(\cdot)), x_{t_{l+1}}(\cdot)) =$$

$$= f(t_l, x(t_l), x_{t_l}(\cdot)) +$$

$$+ \left[\partial_t f(t_l, x(t_l), x_{t_l}(\cdot)) + \frac{\partial f(t_l, x(t_l), x_{t_l}(\cdot))}{\partial x} f(t_l, x(t_l), x_{t_l}(\cdot))\right] \Delta +$$

$$+O(\Delta^2)\,.$$

Substituting these formulas into the residual we obtain $\psi(t_l) = O(\Delta^2)$.

This method has the second convergence order if interpolation and extrapolation of the second order are applied.

To derive the high order coefficients necessary to use the partial and invariant (or coinvariant) derivatives of the corresponding orders and also mixed derivatives of the corresponding high orders. We emphasize that for functionals [8] rules of calculating the coefficients of the Taylor series the exact solutions are the same as for ODE. From this fact, under the condition of sufficient smoothness of the solution, follows the proposition.

Theorem 18.4 *If an explicit Runge-Kutta method for ordinary differential equations has the residual order p, then for functional differential equations the explicit Runge-Kutta-like method with the same coefficients has the same residual order p.*

This statement together with the theorem 18.3[9], allows us construct for functional differential equations analogies

[8] right side of FDE

[9] about the convergence order

of all known for ODE explicit Runge-Kutta methods choosing appropriate methods of interpolation and extrapolation of the discrete model prehistory.

Thus, for example, the improved Euler method with piece-wise linear interppolation (18.2) and extrapolation by continuation (18.8) has the second convergence order. Another *example* is the analog of well-known four-stage ERK-method

$$u_{l+1} = u_l + \frac{1}{6}\Delta\left(h_1 + 2h_2 + 2h_3 + h_4\right),$$

$$h_1 = f(t_l, u_l, u_{t_l}(\cdot)),$$

$$h_2 = f(t_l + \frac{\Delta}{2}, u_l + \frac{\Delta}{2}h_1, u_{t_l+\frac{\Delta}{2}}(\cdot)),$$

$$h_3 = f(t_l + \frac{\Delta}{2}, u_l + \frac{\Delta}{2}h_2, u_{t_l+\frac{\Delta}{2}}(\cdot)),$$

$$h_4 = f(t_l + \Delta, u_l + \Delta h_3, u_{t_l+\Delta}(\cdot)).$$

This method has a fourth convergence order [10] if we apply the interpolation by piece-wise cubic functions and the interpolation by continuation.

For the residual order $p \leq 5$ there are no ERK-like methods with number of stages $k = p$. This fact is known as the *Butcher barrier* [10].

One of ERK-methods with a number of stages $k = 6$ of the order $p = 5$ is the *Runge-Kutta-Fehlberg scheme*, the method composes the basis of standard software of solving ordinary differential equations in well-known software packages MATLAB, MAPLE.

For functional differential equations analog of this method with interpolation by degenerate splines and extrapolation by continuation is realized in Time-delay system toolbox [21][11].

[10]under the conditions of an appropriate smoothness of solutions

[11]see section 24

18.4 Implicit Runge-Kutta-like methods

For functional differential equations, similar to ODE, implicit Runge-Kutta-like methods have some advantages compare to explicit methods:

— for any number of stages k one can construct a method with the convergence order $p = 2k$;

— these methods are suitable for solving stiff systems;

— convenient for parallelization of computational algorithms.

Let an interpolation-extrapolation operator IE with the following properties be fixed:

1) IE operator *consistent*, i.e.

$$u(t_i) = u_i, \ i = l - m, \dots, l,$$

2) IE operator *satisfies the Lipschitz condition*: i.e. there exists a constant L_I such that for any two sets of discrete prehistories $\{u_i^1\}_l$ $\{u_i^2\}_l$ and all $t \in [t_l - \tau, t_l + a\Delta)$ the following inequality is valid

$$\|u^1(t) - u^2(t)\| \leq L_I \max_{l-m \leq i \leq l} \|u_i^1 - u_i^2\|,$$

where $u^1(\cdot) = IE(\{u_i^1\}_l), u^2(\cdot) = IE(\{u_i^2\}_l)$.

The methods of interpolation and extrapolation described in section 18.1, satisfy these conditions.

Definition 18.11 *For a natural k k-stage implicit Runge-Kutta-like method (IRK) with interpolation I and extrapolation E is the numerical model*

$$u_0 = x_0; \tag{18.17}$$

$$u_{l+1} = u_l + \Delta\Psi(t_l, u(\cdot)) =$$

$$= u_l + \Delta \sum_{i=1}^{k} \sigma_i h_i(u_{t_l}(\cdot)), \ l = 1, \dots, N-1, \tag{18.18}$$

$$h_i(u_{t_l}(\cdot)) = f(t_l + a_i\Delta, u_l + \Delta \sum_{j=1}^{k} b_{ij} h_j(u_{t_l}(\cdot)), u_{t_l + a_i\Delta}(\cdot)). \tag{18.19}$$

Here the prehistory of the model is defined by relations: $(\cdot) = IE(\{u_i^1\}_l)$.

Coefficients of the method can be presented by the Butcher matrix

$$\begin{array}{c|cccc} a_1 & b_{11} & b_{12} & \dots & b_{1k} \\ a_2 & b_{21} & b_{22} & \dots & b_{2k} \\ \cdot & \cdot & \cdot & \dots & \cdot \\ a_k & b_{k1} & b_{k2} & \dots & b_{kk} \\ \hline & \sigma_1 & \sigma_2 & \dots & \sigma_k \end{array}$$

Note some special cases:

1) If $b_{ij} = 0$ for $j \geq i$, then we obtain *ERK-like method*;

2) If $b_{ij} = 0$ for $j > i$,then it is *diagonal-implicit method*;

3) If $b_{ij} = 0$ for $j > i$ and $b_{ii} = b_0$, then we obtain the *one-diagonal method.*

Similar to the section 18.2 we denote: $a = \max\{|a_i|, 1 \leq i \leq k\}$, $\sigma = \max\{|\sigma_i|\}$, $b = \max\{|b_{ij}|\}$.

Theorem 18.5 *If*

$$\Delta < \frac{1}{kbL}.$$

Then there exists the unique solution $J = (h_1, h_2, \dots, h_k)'$ of the system (18.19), moreover, the functionals $h_1, h_2, \dots, h_k$, and Ψ satisfy the Lipschitz condition with respect to $u(\cdot)$.

Proof. In the vector space $\mathbb{R}^{kn}$ of the vectors $J = (h_1, h_2, \dots, h_k)'$ consider the norm $||J|| = \max_i ||h_i||$.

The system (18.19) can be written in the form

$$J = \Phi(J).$$

Show that Φ is the constructing operator for a sufficiently small step.

Let $J^1 = (h_1^1, h_2^1, \dots, h_k^1)'$, $J^2 = (h_1^2, h_2^2, \dots, h_k^2)'$, then

$$||\Phi(J^1) - \Phi(J^2)|| =$$

$$= \max_i \|f(t_l + a_i\Delta, u_l + \Delta \sum_{j=1}^{k} b_{ij} h_j^1(u_{t_l}(\cdot)), u_{t_l+a_i\Delta}(\cdot)) -$$

$$-f(t_l + a_i\Delta, u_l + \Delta \sum_{j=1}^{k} b_{ij} h_j^2(u_{t_l}(\cdot)), u_{t_l+a_i\Delta}(\cdot))\| =$$

$$= \|f(t_l + a_{i_0}\Delta, u_l + \Delta \sum_{j=1}^{k} b_{i_0 j} h_j^1(u_{t_l}(\cdot)), u_{t_l+a_{i_0}\Delta}(\cdot)) -$$

$$-f(t_l + a_{i_0}\Delta, u_l + \Delta \sum_{j=1}^{k} b_{i_0 j} h_j^2(u_{t_l}(\cdot)), u_{t_l+a_{i_0}\Delta}(\cdot))\| \leq$$

$$\leq L\Delta b \sum_{j=1}^{k} \|h_j^1 - h_j^2\| \leq L\Delta bk \|J^1 - J^2\|$$

. Therefore, under the accepted assumptions of the step, we have

$$\|\Phi(J^1) - \Phi(J^2)\| \leq \gamma \|J^1 - J^2\|,\ 0 < \gamma < 1.$$

Applying the constructing mapping method to the iteration procedure $J^{i+1} = \Phi(J^i)$, we obtain in the limit the solution of the system (18.19).

h_i satisfies the Lipschitz condition. Indeed let $u^1(\cdot) \in Q[t_l - \tau, t_l + a\Delta)$, $u^2(\cdot) \in Q[t_l - \tau, t_l + a\Delta)$, $u_l^1 = u^1(t_l)$, $u_l^2 = u^2(t_l)$, $h_i^1 = f(t_l + a_i\Delta, u_l^1 + \Delta \sum_{j=1}^{k} b_{ij} h_j^1, u_{t_l+a_i\Delta}^1(\cdot))$, $h_i^2 = f(t_l + a_i\Delta, u_l^2 + +\Delta \sum_{j=1}^{k} b_{ij} h_j^2, u_{t_l+a_i\Delta}^2(\cdot))$, then

$$\|h_i^1 - h_i^2\| \leq L(\|u_l^1 - u_l^2\| + b\Delta \sum_{j=1}^{k} \|h_i^1 - h_i^2\|) +$$

$$+M \sup_{-\tau \leq s < 0} \|u_{t_l+a_i\Delta}^1(s) - u_{t_l+a_i\Delta}^2(s)\| \leq$$

$$\leq b\Delta L \sum_{j=1}^{k} \|h_i^1 - h_i^2\| + (L + M) \sup_{t_l - \tau \leq t < t_l + a\Delta} \|u^1(t) - u^2(t)\|.$$

For the difference of $J^1 = (h_1^1, h_2^1, \ldots, h_k^1)'$ and $J^2 = (h_1^2, h_2^2, \ldots, h_k^2)'$ we have

$$\|J^1 - J^2\| \leq$$

$$\leq bk\Delta L\|J^1 - J^2\| + (L+M) \sup_{t_l - \tau \leq t < t_l + a\Delta} \|u^1(t) - u^2(t)\|$$

or

$$\|h_i^1 - h_i^2\| \leq$$

$$\leq \|J^1 - J^2\| \leq \frac{L+M}{1 - bk\Delta L} \sup_{t_l - \tau \leq t < t_l + a\Delta} \|u^1(t) - u^2(t)\|.$$

The constant $L_J = \frac{L+M}{1-bk\Delta L}$ is a positive number due to the theorem assumptions. From the Lipschitz property it follows that Ψ satisfies the Lipschitz property with respect to $u(\cdot)$ (with the Lipschitz constant $L_\Psi = L_J ka$), i.e.

$$\|\Psi(t_l, u^1(\cdot)) - \Psi(t_l, u^2(\cdot))\| \leq L_\Psi \sup_{t_l - \tau \leq t < t_l + a\Delta} \|u^1(t) - u^2(t)\|.$$

□

Corollary 18.1 *If the interpolation-extrapolation operator IE is fixed, then the function $\Phi(\{u_i\}_l) = \Psi(t_l, IE(\{u_i\}_l))$ satisfies the Lipschitz condition with respect to the discrete model prehistory i.e. there exists a constant $L_\Phi = L_\Psi L_I$ such that the inequality*

$$\|\Phi(\{u_i^1\}_l) - \Phi(\{u_i^2\}_l)\| \leq L_\Phi \max_{l-m \leq i \leq l} \|u_i^1 - u_i^2\|.$$

is held.

A residual order and a convergence order of an IRK-like method are defined similarly to the ERK-like methods

Theorem 18.6 *If the method (18.17) - (18.19) has the residual order $p_1 > 0$ and, the interpolation-extrapolation operator of the model prehistory has the order $p_2 > 0$, then the method converges, moreover, the convergence order p of the IRK-like method not less than the minimum of p_1, p_2.*

The theorem can be proved similar to the theorem 18.4.

Remark 18.1 For FDE, similar to ODE, for any number of stages k one can construct the method which will have a convergence order of $p = 2k$.

For $k = 1$, the corresponding method is the implicit rule of the middle point IRK-like method with the Butcher matrix

$$\begin{array}{c|c} \frac{1}{2} & \frac{1}{2} \\ \hline & 1 \end{array}.$$

For $k = 2$ the corresponding scheme is the Hammer-Hollingsworth method of the order 4:

$$\begin{array}{c|cc} \frac{1}{2} - \frac{\sqrt{3}}{6} & \frac{1}{4} & \frac{1}{4} - \frac{\sqrt{3}}{6} \\ \frac{1}{2} + \frac{\sqrt{3}}{6} & \frac{1}{4} + \frac{\sqrt{3}}{6} & \frac{1}{4} \\ \hline & \frac{1}{2} & \frac{1}{2} \end{array}.$$

Remark 18.2 For realizing IRK-like methods at every step, it is necessary to solve the system (18.18) with respect to u_{l+1} (or system (18.19) with respect to h_i). The following realizations are possible the:

1) Predict-corrector scheme.: Make a prediction u^0_{l+1} by an explicit method, then make one iteration by the method of simple iteration

$$u^1_{l+1} = u^1_l + \Delta \sum_{i=1}^{k} \sigma_i \, h_i(u^0_{t_l}(\cdot)).$$

2)Scheme $P(EC)^m$: Prediction is made one time, and computation-correction — m times:

$$u^{i+1}_{l+1} = u^m_l + \Delta \sum_{i=1}^{k} \sigma_i \, h_i(u^i_{t_l}(\cdot)), \quad i = 0, ..., m-1.$$

A variant of this scheme is the $P(EC)^m E$ scheme, when the value of the functional in the right-hand side is calculated additionally one time for utilizing at the next step.

3) The gradient Newton-Raphson method for solving nonlinear systems. The main computational problem of the method is the calculation of the Jacobi matrix

Remark 18.3 IRK-methods for ODE are the main methods of solving stiff problem. [13, 12].For FDE this aspect is not well studied yet.

Remark 18.4 *Potential for parallel computation .*

Implicit methods (18.17) – (18.19) have considerable potential for parallel computations. It is known that for ODE there are three basic directions of the parallelization of the Runge-Kutta methods:

1) To use IRK-methods in which the Butcher matrices allows one to simultaneously apply several processors for calculating functionals h_i by formulas (18.19). For example, the 4-stage method

$$\begin{array}{c|cccc} \frac{1}{2} & \frac{1}{2} & 0 & 0 & 0 \\ \frac{2}{3} & 0 & \frac{2}{3} & 0 & 0 \\ \frac{1}{2} & -\frac{5}{2} & \frac{5}{2} & \frac{1}{2} & 0 \\ \frac{1}{3} & -\frac{5}{3} & \frac{4}{3} & 0 & \frac{2}{3} \\ \hline & -1 & \frac{3}{2} & -1 & \frac{3}{2} \end{array}$$

allows to use 2 processors simultaneously for calculating h_1 & h_2, and then h_3 & h_4.

2) Parallelism in *computing algebra* is, first of all in the factorization of matrices of the Newton-Raphson method: to find h_i it is necessary to solve nonlinear equations

(18.19). For numerical solutions of such systems, the Newton-Raphson method is usually applied, however every iteration of this method consists of solving a linear algebraic equations of a large dimension. Methods of factorization of the Jacobi matrix, which are usually applied in ODE case for parallelization of computation, can be also applied in FDE case.

3) Similar to ODE one can use parallelization for computation of the coordinates of functionals $f(t_n, u_n, u_{t_n}(\cdot))$ in (23.1).

For FDE, in addition to these possibilities for parallelization, one can use the following:

4) Parallelization of calculating interpolation and extrapolation of all coordinates of the discrete model prehistory.

5) Parallelization of methods of calculating integrals which describe distributed delays in right-hand sides of systems with delays (as rule FDE with *distributed* delays are described by means of *integrals*, and during the calculation the most of time is spent for calculating these integrals.)

6) Parallelization of the computation of *several* integrals for FDE containing a several *distributed* delays.

19 Multistep numerical methods

Multistep numerical methods based on approximation of functionals [12] along solutions for were elaborated by Kemper & Tavernini. In [23, 24] another approach (based on interpolation of the discrete models) is realized. In this

[12] in the right-hand part of FDE

section we follow this approach.

19.1 Numerical models

It is supposed that a partition $\{t_n\}_{-m}^{N}$ of the interval $[t_0 - \tau, \theta]$, and an interpolation-extrapolation operator

$$IE : \{u_i\}_n \to u(\cdot) \in Q[t_n - \tau, t_n + \Delta]$$

are fixed.

An *explicit k-step method* is the discrete model

$$u_n = \sum_{i=1}^{k} \alpha_i \, u_{n-i} + \Delta \sum_{i=1}^{k} \beta_i \, f_{n-i}^*, \quad n = k, \ldots, N, \qquad (19.1)$$

where α_i, β_i are the parameters of the method and $f_i^* = f(t_i, u_i, u_{t_i}(\cdot))$, $i = 1, \ldots, k$.

The *implicit k-step method* is the discrete model

$$u_n = \sum_{i=1}^{k} \alpha_i \, u_{n-i} + \Delta \sum_{i=0}^{k} \beta_i \, f_{n-i}^*. \qquad (19.2)$$

It is necessary to note that we can apply discrete models (19.1) and (19.2) for $n = k, \ldots$ only if we know *starting values* u_i, $i = 0, \ldots, k-1$, which define f_i^*, $i = 0, \ldots, k-1$. So we say that we have a *starting procedure of an order p* if there exists a constant $C > 0$ such that $\|x_i - u_i\| \leq C\,\Delta^p$ as $i = 0, \ldots, k-1$.

Also note that an extrapolation is necessary only for implicit multistep methods.

19.2 Order of convergence

In order to obtain error estimates of a method let us introduce the variable $z_n = u_n - x_n$ and substitute $u_{n-i} =$

$x_{n-i} + z_{n-i}$, $i = 0, \dots, k$, into (19.2). We obtain

$$z_n = \sum_{i=1}^{k} \alpha_i \, z_{n-i} + \Delta \, \psi_n^{(1)} + \Delta \, \psi_n^{(2)},$$

where

$$\psi_n^{(1)} = \frac{1}{\Delta} \left(\sum_{i=1}^{k} \alpha_i \, x_{n-i} - x_n \right) + \sum_{i=0}^{k} \beta_i \, f_{n-i} \,, \qquad (19.3)$$

$$f_i = f(t_i, x(t_i), x_{t_i}(\cdot)) \,,$$

$$\psi_n^{(2)} = \sum_{i=0}^{k} \beta_i (f_{n-i}^* - f_{n-i}) \,.$$

The value $\psi_n^{(1)}$ is called an *approximation error* (residual) of a multistep method.

Definition 19.1 *A multistep method has an* approxima*tion error of* order p *if there exists a constant C such that*

$$\|\psi_n^{(1)}\| \leq C \Delta^p \quad \textit{for all} \quad n = k, \dots, N.$$

□

Definition 19.2 *Method (19.2)*
1) Converges, if $\|z_n\| \to 0$ as $\Delta \to 0$ for all $n = 0, \dots, N$;
2) Has a convergence order p *if there exists a constant C such that*

$$\|z_n\| \leq C \Delta^p \quad \textit{for all} \quad n = k, \dots, N;$$

3) Is 0-stable *if the* root condition*[7] is fulfilled, i.e., all roots ξ of the generating polynomial*

$$\rho^k - \sum_{i=1}^{k} \alpha_i \, \rho^{n-i}$$

satisfy the inequality $|\xi| \leq 1$; if $|\xi| = 1$, then ξ is the prime root. □

The following proposition is true.

Theorem 19.1 *If the multistep method (19.2) is 0-stable and has*
1) An approximation error of order p_1,
2) A starting procedure error of an order p_2,
3) An error of prehistory interpolation and extrapolation of order p_3.
Then the method converges and has the convergence order $p \geq \min\{\, p_1,\, p_2,\, p_3 \,\}$.

Proof. The proof of the theorem repeats [13] the proof of the corresponding theorem of ODE case. So we remain the proof to a reader as an exercise on application of i-Smooth analysis to FDE numerical methods. □

19.3 Approximation order. Starting procedure

Let the solution of the initial value problem (23.1) – (23.3) be $(p+1)$-times continuous differentiable on $[t_0, t_0 + \theta]$. Then from the Taylor formula we have

$$x_{n-i} = \sum_{j=0}^{p} \frac{(-i\Delta)^j\, x^{(j)}(t_n)}{j!} + O(\Delta^{p+1}),$$

$$f_{n-i} = \sum_{j=0}^{p} \frac{(-i\Delta)^j\, x^{(j+1)}(t_n)}{j!} + O(\Delta^{p}).$$

Substituting these formulas into (19.3) one can find that an order of a residual of the method greater or equal to $p \geq 1$ if the following conditions are satisfied

$$\sum_{i=1}^{k} a_i = 1,$$

[13]with obvious application of the invariant derivatives in appropriate places

$$\sum_{i=0}^{k} i^{j-1}\,(j\,\beta_i - i\,\alpha_i) = 0\ ,\ \ j = 1, \ldots, p\,.$$

Hence an order p of an approximation error satisfies the inequality $p \leq 2k$ for *implicit methods* and the inequality $p \leq 2k - 1$ for *explicit methods.*

Similar to ordinary differential equations [10], if k-step method is 0-stable then the following inequalities are fulfilled :

1) $p \leq k + 2$ for even k,
2) $p \leq k + 1$ for odd k,
3) $p \leq k$ for $\beta_0 \leq 0$ (in particular, for explicit methods).

The approximation error orders of the Adams methods

$$u_n = u_{n-1} + \Delta\left(\sum_{i=0}^{k} \beta_i\, f^*_{n-i}\right)$$

satisfy the inequality $p \leq k + 1$ for implicit methods and the inequality $p \leq k$ for explicit methods.

The starting procedures for k-step method can be realized on the basis of Runge–Kutta method of the corresponding approximation order.

Remark 19.1 T.S. Zverkina constructed [39] for some classes of FDE the modified Adams formulas

$$u_n = u_{n-1} + \Delta\left(\sum_{i=0}^{k} \beta_i(\gamma)\, f^*_{n-i}\right),$$

which can handle discontinuities of a solution derivatives. □

20 Startingless multistep methods

It is necessary to organize a *starting procedure* for the realization of numerical methods (19.1) and (19.2). Below we discuss *startingless* multistep numerical methods.

20.1 Explicit methods

If $m \geq k$ then an *explicit startingless k-step method* is the numerical model

$$u_n = \sum_{i=1}^{k} \alpha_i \, u_{n-i} + \Delta \, \beta_1 \, f^*_{n-1} \,, \quad n = 1, 2, \ldots, N, \qquad (20.1)$$

where

$$u_0 = x^0 \,,$$

$$u_i = x_i \equiv y^0(t_i - t_0) \quad \text{as} \quad -m \leq i < 0.$$

From the results of the section 16 it follows that an order p of a residual of the method (20.1) satisfies inequality $p \leq k$.

Two following 0-stable methods of the form (20.1) have $p = k$:
the explicit Euler method

$$u_n = u_{n-1} + \Delta \, f^*_{n-1} \,, \quad p = k = 1 \,;$$

and the middle point method

$$u_n = u_{n-2} + 2 \, \Delta \, f^*_{n-1} \,, \quad p = k = 2 \,.$$

For $k \geq 3$ there are no 0-stable methods (20.1) with $p = k$. One can construct 0-stable 4-step startingless methods, which have the approximation order of 3. The following formula gives an example of such methods

$$u_n = -\frac{2}{5} \, u_{n-1} + \frac{6}{5} \, u_{n-2} + \frac{2}{5} \, u_{n-3} - \frac{1}{5} \, u_{n-4} + \frac{12}{5} \, \Delta \, f^*_{n-1}$$

has $k = 4$, $p = 3$.

20.2 Implicit methods

For $m \geq k$ an *implicit startingless k-step method* is the numerical model

$$u_n = \sum_{i=1}^{k} \alpha_i \, u_{n-i} + \Delta \left(\beta_0 \, f_n^* + \beta_1 \, f_{n-1}^* \right) , \quad n = 1, \ldots, N, \tag{20.2}$$

with the initial discrete prehistory

$$u_0 = x^0 ,$$
$$u_n = y^0(t_n - t_0) , \; n = -m, \ldots, -1.$$

The approximation order p of the method (20.2) satisfies the inequality $p \leq k + 1$.

For $p = k + 1$ only the trapezoidal method

$$u_n = u_{n-1} + \frac{\Delta}{2}(f_n^* + f_{n-1}^*), \quad k = 1, \; p = 2$$

is 0-stable.

Among methods (20.2) with $p = k$, we note the formulas of backwards differentiation (FBD)

$$\sum_{i=1}^{k} \frac{1}{i} \, \nabla^i \, u_n = \Delta \, f_n^*$$

(here ∇ denote the backwards difference: $\nabla^0 u_n = u_n$, $\nabla^{i+1} u_n = \nabla^i u_n - \nabla^i u_{n-1}$) which are 0-stable for $k \leq 6$ [10]. For example, a 2-step method of the second order is

$$\frac{3u_n - 4u_{n-1} + u_{n-2}}{2} = \Delta \, f_n^* .$$

These methods can be applied for solving stiff problems.

If the exact solution $x(t)$ of the problem (0.1) - (0.3) is (p+1)-time continuous differentiable on the interval $[t_0, t_0 + \theta]$ then from the Taylor formula we have

$$x_{l-i} = \sum_{j=0}^{p} \frac{(-i\Delta)^j \, x^{(j)}(t_l)}{j!} + O(\Delta^{p+1}) ,$$

$$f_{l-i} = \sum_{j=0}^{p} \frac{(-i\Delta)^j \, x^{(j+1)}(t_l)}{j!} + O(\Delta^p) \, .$$

Substituting these expressions into the residual (19.3) we obtain that the residual order of the method (19.2) is not less than $p \geq 1$, if the following conditions are satisfied

$$\sum_{i=1}^{k} \alpha_i = 1 \, ,$$

$$\sum_{i=0}^{k} i^{j-1} \, (j \, \beta_i - i \, \alpha_i) = 0 \ , \ j = 1, \ldots, p \, .$$

Therefore the approximation order of implicit methods is $p \leq 2k$, and for explicit methods: $p \leq 2k - 1$.

However, similar to ODE for 0-stable k-stage k-step methods the *Dahlqist barriers* hold [10]:
1) $p \leq k + 2$ for even numbers k,
2) $p \leq k + 1$ for odd numbers k,
3) $p \leq k$ if $\beta_0 \leq 0$ (in particular for explicit methods).

For *Adams methods*

$$u_l = u_{l-1} + \Delta \left(\sum_{i=0}^{k} \beta_i \, f^*_{l-i} \right)$$

the order of the approximation error $p \leq k + 1$ for implicit methods and $p \leq k$ for explicit methods.

So for a general class of multistep methods one can achieve great accuracy for implicit methods but only for even numbers k.

Remark 20.1 For the realization of implicit methods at every step it is necessary to solve the system of equations (19.2) with respect to $u_{(l)}$. There following realizations are possible

1) The *predict-correct scheme.* Make a prediction u_l^0 by an explicit method, then one iteration by the method of

simple iteration

$$u_l^1 = \sum_{i=1}^{k} \alpha_i\, u_{l-i} + \Delta \sum_{i=1}^{k} \beta_i\, f^*_{l-i} + \Delta\beta_0 f(t_l, u_l^0, u_{t_l}(\cdot)).$$

2) *Scheme* $P(EC)^m$, i.e the prediction is made just one time and corrections are made m times:

$$u_l^{i+1} = \sum_{i=1}^{k} \alpha_i\, u_{l-i} + \Delta \sum_{i=1}^{k} \beta_i\, f^*_{l-i} + \Delta\beta_0 f(t_l, u_l^i, u_{t_l}(\cdot)),$$

$$i = 0, ..., m-1.$$

A variant of this scheme is the *scheme* $P(EC)^m E$, in which the value of the functional in the right-hand side of the system is calculated additionally one time for utilization at the next step.

3) *Gradient method* (Newton-Raphson method) of solving the nonlinear system (19.2). Main computational problem of the method is the calculation of the Jacobi matrix.

20.3 Startingless multistep methods

A specific feature of FDE is that from (0.2) - (0.3) only model values u_i for $-m \le i \le 0$ can be derived, whereas (while) formulas (19.1) or (19.2) can be applied only with a starting procedure(i.e. for $l = 1, 2, \ldots, k-1$), because the values f_i^* for $i < 0$ are not defined.

Definition 20.1 *For $m \ge k$, the explicit k-step startingless method, is the numerical model*

$$u_l = \sum_{i=1}^{k} \alpha_i\, u_{l-i} + \Delta\, \beta_1\, f^*_{l-1}\,, \quad l = 1, 2, \ldots, N, \tag{20.3}$$

where

$$u_i = x_i \equiv y^0(t_i - t_0) \ \ for \ \ -m \le i < 0,$$

$$u_0 = x_0.$$

From results of the previous section it follows that the residual order p of the method (20.3) satisfies the condition $p \leq k$. 0-stable methods (20.3), which have $p = k$, are:
the explicit Euler method

$$u_l = u_{l-1} + \Delta\, f^*_{l-1}\,, \quad p = k = 1,$$

and the middle point rule

$$u_l = u_{l-2} + 2\,\Delta\, f^*_{l-1}\,, \quad p = k = 2\,.$$

there are no 0-stable among methods of the form (20.3) for $k \geq 3$ with $p = k$. However are explicit four-step 0-stable startingless methods of the third approximation order, for example

$$u_l = -\frac{2}{5}\, u_{l-1} + \frac{6}{5}\, u_{l-2} + \frac{2}{5}\, u_{l-3} - \frac{1}{5}\, u_{l-4} + \frac{12}{5}\, \Delta\, f^*_{l-1}.$$

For this method $k = 4$, $p = 3$.

Definition 20.2 *For $m \geq k$ the implicit k-step startingless method is the model*

$$u_l = \sum_{i=1}^{k} \alpha_i\, u_{l-i} + \Delta\,(\beta_0\, f^*_l + \beta_1\, f^*_{l-1})\,, \; l = 1, \ldots, N. \quad (20.4)$$

The residual order p of the method (20.4) satisfies the condition $p \leq k + 1$.

Only the trapezoidal method

$$u_l = u_{l-1} + \frac{\Delta}{2}(f^*_l + f^*_{l-1}), \quad k = 1,\; p = 2.$$

is 0-stable among methods (20.4) with $p = k + 1$. Among methods (20.4) with $p = k$, note the *backwards differentiation formula* (BDF)

$$\sum_{i=1}^{k} \frac{1}{i}\, \nabla^i\, u_l = \Delta\, f^*_l$$

(here ∇ denotes the backwards difference: $\nabla^0 u_l = u_l$, $\nabla^{i+1} u_l = \nabla^i u_l - \nabla^i u_{l-1}$), 0-stable for $k \leq 6$ [10]. These methods can be applied are applied for solving stiff systems. As an example two-step method of the second order has the form

$$\frac{3u_l - 4u_{l-1} + u_{l-2}}{2} = \Delta f_l^* .$$

21 Nordsik methods

Nordsik methods [10] are one of the most economic for ODE. They are equivalent to multistep methods, however it is easily to change the step size in these methods.

Coefficients of the Nordsik methods approximate derivatives of the exact solution, but the derivatives are calculated just one time for obtaining starting values. In case of FDE we apply i-smooth calculus for calculating derivatives.

For simplicity in this section we consider only one-dimensional FDE [14] (0.1), i.e. $x \in \mathbb{R}$.

Similar to previous sections we consider the grid and an interpolation-extrapolation operator

$$IE : \{u_i\}_l \to u(\cdot) \in Q[t_l - \tau, t_l + \Delta]' .$$

We introduce the vector $U_l \in \mathbb{R}^k$ *(Nordsik vector)* with the coordinates $U_l = (u_l^1, u_l^2, \ldots, u_l^k)$, $u_l^1 = u_l$ approximates the exact solution $x(t_l)$, the second coordinates u_l^2 approximates $\Delta \dot{x}(t_l)$, and so on, the last coordinate u_l^k approximates $\frac{\Delta^{k-1}}{(k-1)!} x^{(k-1)}(t_l)$. Here $l = 0, \ldots, N$.

The *Nordsik method* is the numerical model

$$U_{l+1} = PU_l + q[\Delta f(t_{l+1}, u_{l+1}, u_{t_{l+1}}(\cdot)) - e_2' P u_l], \quad (21.1)$$

$$l = 0, \ldots, N-1,$$

[14]we do not realize these methods in our software package presented in the next chapter

where P is the triangle Pascal matrix of the dimension $k \times k$, in which elements P_{ij} are defined by formulas:

$$P_{ij} = \begin{cases} C_j^i, & \text{for } 1 \leq i \leq j \leq k, \\ 0, & \text{otherwise,} \end{cases}$$

where C_j^i is i-fold combinations of j elements; the vector e_2 is defined by the relation $e_2 = (0, 1, 0, \dots, 0)'$; coordinates of the vector $q = (q_1, 1, q_3, \dots, q_l)'$ are parameters of the vector.

The condition $q_2 = 1$ defines relations of the method with the equation (0.1): from the step-by-step formula it follows $u_{l+1}^2 = f(t_{l+1}, u_{l+1}, u_{t_{l+1}}(\cdot))$.

Introducing the matrix $S = P - qe_2'P$, we can present a step-by step formula in the form

$$U_{l+1} = SU_l + q\Delta f(t_{l+1}, u_{l+1}, u_{t_{l+1}}(\cdot)), \; l = 0, \dots, N-1. \tag{21.2}$$

Numbers 1 and 0 are the eigenvalues of the matrix S, and its characteristic polynomial does not depend on the parameter q_1.. One can find [30]with such numbers $q_3, \dots, q_k$ that other eigenvalues would be equal to the zero. In this case the method is 0-stable (in the sense of definitions of the section 19.2). The coefficient q_1 can be chosen from the condition that the constant of the approximation equals to zero [10, p. 384]. For example for $k = 4$ the matrix S has the form

$$S = \begin{pmatrix} 1 & 1-q_1 & 1-2q_1 & 1-3q_1 \\ 0 & 0 & 0 & 0 \\ 0 & -q_3 & 1-2q_3 & 3-3q_3 \\ 0 & -q_4 & -2q_4 & 1-3q_3 \end{pmatrix},$$

and the vector $q = \left(\frac{3}{8}, 1, \frac{3}{4}, \frac{1}{6}\right)$. The starting values are defined by formulas

$$u_0^1 = x(t_0) = x_0,$$

$$u_0^2 = \Delta \dot{x}(t_0) = \Delta f(t_0, x_0, y^0(\cdot)),$$

$$\cdot \quad \cdot \quad \cdot$$

$$u_0^k = \frac{\Delta^{k-1}}{(k-1)!} x^{(k-1)}(t_0).$$

It is convenient to use the technique of i-smooth analysis for calculating the derivatives of functionals in the right-hand sides of (0.1)

$$u_0^3 = \frac{\Delta^2}{2} \frac{d}{dt} [f(t, x, y(\cdot))]|_{t=t_0, x=x_0, y=y^0} =$$

$$= \frac{\Delta^2}{2} [\partial_t f(t_0, x_0, y^0(\cdot)) + \frac{\partial f(t_0, x_0, y^0 \cdot))}{\partial x} f(t_0, x_0, y^0(\cdot))].$$

Define the discrete *function of the exact values* – the vector $Z_l \in \mathbb{R}^k$ with coordinates $Z_l = (z_l^1, z_l^2, \cdots, z_l^k)$ by relations ($l = 0, \dots, N$)

$$z_l^1 = x(t_l), \; z_l^j = \frac{\Delta^{j-1}}{(j-1)!} x^{(j-1)}(t_l), \; j = 2, \dots, k.$$

The *residual* (approximation error) of the Nordsik method is the function

$$\psi(t_l) = \frac{Z_{l+1} - SZ_l}{\Delta} - qf(t_{l+1}, x_{l+1}, x_{t_{l+1}}(\cdot)), \; l = 0, \dots, N-1.$$

The residual has an order p, if there exists a constant C such that $\|\psi(t_l)\| \le C\Delta^p$ for all $l = 0, 1, \dots, N-1$.

The method (21.2) *converges*, if $\|Z_l - U_l\| \to 0$ as $\Delta \to 0$ for all $l = 0, \dots, N$, and has a *convergence order* p, if there exists C such that $\|Z_l - U_l\| \le C\Delta^p$ for all $l = k, \dots, N$.

Theorem 21.1 *Let in the Nordsik method (21.2), coordinates of the vector q are chosen in a way of guaranteeing 0-stability. Besides this, the method has the residual of the order p_1 and the order of the interpolation and the extrapolation of the discrete model prehistory is equal to p_3. Then the method converges and has the convergence order $p \ge \min\{p_1, p_2\}$.*

Proof of the theorem follows from proposition [10, p. 382].

It also follows from the fact that the Nordsik method is equivalent to multistep methods and also from the theorem 17.1 and the inclusion of the Nordsik method into the general scheme of the section 22.

Remark 21.1 It is easy to change the order of the multistep methods. It is necessary only add the next term into the presentation (19.2). However the change of the step is a sufficiently complicated procedure. In the Nordsik presentation(21.1) it is vise versa, a change of the order is not natural, however one can easily change the step by multiplying the Nordsik vector by a diagonal matrix with diagonal elements $(1, \omega, \omega^2, \ldots)$, where $\omega = h_{new}/h_{old}$ is the ratio of steps.

21.1 Methods based on calculation of high order derivatives

One can derive a method of arbitrary approximation order for a concrete FDE (0.1) if using the Taylor expansion of the exact solution $x(t)$ or the functional f. For example the formula

$$x(t_{l+1}) \approx x(t_l) + \Delta\dot{x}(t_l) + \frac{\Delta}{2}\ddot{x}(t_l)$$

with application of rules calculating derivatives of the functional along with solutions

$$\ddot{x}(t) = \frac{d}{dt}f(t, x(t), x_t(\cdot)) =$$

$$= \partial_t f(t, x(t), x_t(\cdot)) + \frac{\partial f(t, x(t), x_t(\cdot))}{\partial x} f(t, x(t), x_t(\cdot))$$

gives (in combination with a fixed operator $I : \{u_i\}_l \rightarrow u(\cdot) \in Q[t_l - \tau, t_l]$ of the model prehistory interpolation) the method

$$u_{l+1} = u_l + \Delta f(t_l, u_l, u_{t_l}(\cdot)) +$$

$$+\frac{\Delta^2}{2}\left[\partial_t f(t_l, u_l, u_{t_l}(\cdot)) + \frac{\partial f(t_l, u_l, u_{t_l}(\cdot))}{\partial x} f(t_l, u_l, u_{t_l}(\cdot))\right]. \tag{21.3}$$

In contrast to the Runge-Kutta-like methods, the method does not require the extrapolation of the model prehistory. This method has the second residual order that directly follows from the Taylor expansion, so it is reasonable to use piece-wise linear interpolation.

The convergence order in this case is equal to 2, that can be proved similarly to theorem 18.3. Such methods are convenient to apply when the structure of the functional in the right-hand side of FDE (0.1) is fixed.

Example 21.1 Consider the linear system with a constant discrete and distributed delays

$$\dot{x}(t) = Ax(t) + Bx(t-\tau) + \int_{-\tau}^{0} G(s)x(t+s)ds. \tag{21.4}$$

Applying the technique of i-smooth calculus one can calculate the gradient and the coinvariant derivative [15] of the functional

$$f(t, x, y(\cdot)) = Ax + By(-\tau) + + \int_{-\tau}^{0} G(s)y(s)ds$$

which is the right-hand side of system (21.4):

$$\frac{\partial f}{\partial x} = A;$$

$$\partial_t f = B\dot{x}(t-\tau) + G(0)x(t) - G(-\tau)x(t-\tau).$$

If denote $l = 0, \dots, N$, $N = \theta/\Delta$, $m = \tau/\Delta$)

$$f_l = f(t_l, u_l, u_{t_l}(\cdot)) = Au_l + Bu_{l-m} + \int_{-\tau}^{0} G(s)u(t_l+s)ds, \tag{21.5}$$

[15] in this specific case coinvariant derivative is equal to the invariant derivative

then the method (21.3) has the form

$$u_{l+1} = u_l + \Delta f_l + \frac{\Delta^2}{2}\left[B f_{l-m} + G(0)u_l - G(-\tau)u_{l-m} + A f_l\right]. \tag{21.6}$$

In a similar way one can use a higher order.

System (0.1) can be written in the coordinate form

$$\dot{x}_i(t) = f_i(t, x(t), x_t(\cdot)),\ i = 1, ..., n,$$

then

$$\frac{d^2}{dt^2}(x_i(t)) = \partial_t f_i(t, x(t), x_t(\cdot)) + \frac{\partial f_i(t, x(t), x_t(\cdot))}{\partial x_1} f_1(t, x(t), x_t(\cdot)) +$$

$$+ \cdots + \frac{\partial f_i(t, x(t), x_t(\cdot))}{\partial x_n} f_n(t, x(t), x_t(\cdot)). \tag{21.7}$$

If a functional $\Psi(t, x(t), x_t(\cdot)) : [t_0, t_0 + \theta] \times \mathbb{R}^n \times Q[-\tau, 0) \to \mathbb{R}$ has appropriate properties of smoothness then by D we can denote the differential operator acting on the functional Ψ by the rule

$$(D\Psi)(t, x(t), x_t(\cdot)) = \partial_t \Psi(t, x(t), x_t(\cdot)) +$$

$$+ \frac{\partial \Psi(t, x(t), x_t(\cdot))}{\partial x_1} f_1(t, x(t), x_t(\cdot)) +$$

$$+ \ldots + \frac{\partial \Psi(t, x(t), x_t(\cdot))}{\partial x_n} f_n(t, x(t), x_t(\cdot)).$$

Because $Dx_i = f_i$, therefore we obtain formulas [16]:

$$\frac{d^j}{dt^j}(x_i(t)) = (D^j x_i)(t, x(t), x_t(\cdot)).$$

This notation is convenient for defining a new class of numerical methods posessing as well as specifics of Runge-kutta-like methods, and methods utilizing the Taylor expansion.

[16] which generalize(21.7)

For a natural k *k-stage explicit Runge-Kutta-like method with application of derivatives of order* q and interpolation-extrapolation operator IE is the numerical model

$$u_{l+1} = u_l + \sum_{r=1}^{q} \sum_{j=1}^{k} \sigma_j^r h_j^r(u_l, u_{t_l}(\cdot)),\ l = 1, ..., N-1, \quad (21.8)$$

$$h_i^n(u_l, u_{t_l}(\cdot)) = \frac{\Delta^n}{n!}(D^n x)\cdot$$

$$\cdot(t_l + a_i\Delta, u_l + \sum_{r=1}^{q} \sum_{j=1}^{k} b_{ij}^r h_j^r(u_l, u_{t_l}(\cdot)), u_{t_l+a_i\Delta}(\cdot)).$$

Here the model prehistory $u_t(\cdot)$ is defined by the operator IE.

Numbers

$$a_i,\ i = 1, \ldots, k,$$

$$\sigma_j^r,\ j = 1, \ldots, k,\ r = 1, \ldots, q,$$

$$b_{ij}^r,\ i, j = 1, \ldots, k,\ r = 1, \ldots, q$$

are the coefficients of the method.

This is the explicit method if all b_{ij}^r equal to zero for $i \le j$, otherwise this is the implicit method.

These methods are equivalent to *collocational* methods with multiple knots and include the Obreshkov methods, the Fehlberg method and other methods [10].

21.2 Various methods based on the separation of finite-dimensional and infinite-dimensional components of the phase state

Our approach to constructing numerical methods for FDE is based on the methodology of separating finite-dimensional and infinite-dimensional components in the phase structure of the system (0.1). This methodology together with the interpolation of the discrete model prehistory allows us to elaborate for the functional differential equations complete analogies of all that is known for ODE numerical algorithms.

Note the following basic types of numerical methods which are elaborated for FDE [17]:

1) One-step k-stage (Runge-Kutta-like methods),

2) Multistep methods,

3) Methods utilizing calculation of high order derivatives,

4) Extrapolation methods,

5) Cyclic methods,

6) Block methods,

7) Hybrid methods,

8) Multistep multistage methods applying the calculation of high order derivatives,

9) Nordsik methods,

10) Methods of solving second order equations,

11) Simplicial (simplectic) methods.

Note these are *positional* methods, i.e. at a moment t one can define a part of trajectory till the moment, in contradiction to, for example, methods solving FDE by reducing to integral equation, where the solution is defined on the whole interval.

Extrapolation methods

In section 23.3 we describe for FDE the Runge rule of estimating the approximation order and the Richardson extrapolation in case of a double step decreasing. This process can be continued by decreasing the step in 2, 4, 8, 16,... times. The *Bulirsh sequence* 1, 2, 3, 4, 6, 8, 12, 16, 24,... is the most economic sequence [10]. Extrapolation method have advantages that at every step one can change both the step length and the order of a method due to which a high accuracy can be archived. Theoretical basis of these methods is the asymptotic representation of the global er-

[17]in this book we describe only the most interesting for the author

ror. This result will be presented in the section 22.4 for FDE case.

Cyclic methods

Applying cyclically several different multistep methods [10]. one can obtain converging methods of order for $2k-1$ (or $2k$ for even k). This destroys the *Dahlkvist barrier* (see section 19.3).

Block methods

By the Runge-Kutta-like methods, once the block is u_{l+1}. The block methods require less than the Runge-Kutta-like methods calculations of values f.

Hybrid methods

In linear multistep methods with information at points $t_l = t_0 + l\Delta$, the hybrid methods unify properties of the multistep methods and the Runge-Kutta-like methods. For example k-stage hybrid formula is

$$\sum_{i=0}^{k} \alpha_i u_{l+i} = \Delta \sum_{i=0}^{k} \beta_i f_{l+i} + \Delta \beta_\nu f_{l+\nu},$$

where ν is a natural. An appropriate choice of the parameters ν and β_ν allows one to obtain numerical methods with better properties that have the multistep and the Runge-Kutta-like methods [13].

Multistep multistage methods with the application of higher derivatives

Apparently these methods compose the widest class of explicit form linear methods, as was noted in the section 21.1.

These methods can be presented in the form:

$$u_{l+1}^i = \sum_{j=1}^{k} a_{ij} u_l^j + \sum_{r=1}^{q} \frac{\Delta^r}{r!} \sum_{j=1}^{s} bij^r D^r x(t_l + c_j \Delta, v_l^j, v_{t_l+c_j\Delta}^j(\cdot)),$$

$$i = 1, \ldots, k,$$

$$v_l^i = \sum_{j=1}^{k} \hat{a}_{ij} u_l^j + + \sum_{r=1}^{q} \frac{\Delta^r}{r!} \sum_{j=1}^{s} \hat{b}ij^r D^r x(t_l + c_j \Delta, v_l^j, v_{t_l+c_j\Delta}^j(\cdot)),$$

$$i = 1, \ldots, s.$$

Stages (u_l^i, $i = 1, \cdots, k$) are called the *external stages* and contain information necessary for the next stage, stages (v_l^i, $i = 1, \cdots, s$) are called the *interior stages.* It is assumed that an interpolation-extrapolation operator

$$IE : \{u_i\}_l \to v(\cdot) \in Q[t_l - \tau, t_l + c\Delta]),$$

with $c = \max_i \{c_i\}$ is fixed.

Methods of solving second order equations

The second order differential equations

$$\ddot{x} = f(t, x, \dot{x})$$

are the usual models in mechanics. Methods elaborated for solving such equations are possibly the oldest among numerical methods. Nevertheless they still do not lose their certainty because they effectively take into account specific features of the object under consideration. One can separate these methods into two main classes:

– one step multistage methods (for example, the Nustrem method [10]).

– multistep methods (for example, the Steiner method [2]).

Similar methods can be developed for numerical solving of functional differential equations

$$\ddot{x}(t) = f(t, x(t), \dot{x}(t), x_t(\cdot), \dot{x}_t(\cdot)).$$

Simplectic methods

Methods conserving the phase volume are called Liouville methods. These methods can be applied for calculating closed trajectories (cycles). An example of such method is a pair: explicit and implicit Euler methods which are applied in turns. Close to the simplectic methods are the *projection* methods.

These and other methods are included into a general scheme describing in the following section.

22 General linear methods of numerical solving functional differential equations

22.1 Introduction

Consider the initial value problem

$$\dot{x}(t) = f(t, x_t(\cdot)), \quad t \in [\alpha, \beta], \tag{22.1}$$

$$x(t) = \phi(t), \quad t \in [\alpha - \tau, \alpha]. \tag{22.2}$$

Here $\tau \geq 0$ is the delay, $x \in \mathbb{R}^n$ is the phase vector, $\mathbb{R}^n$ is n-dimensional Euclidean space with the norm$\| \cdot \|$, $x_t(\cdot) = \{x(t+s), \alpha - \tau \leq t+s \leq t\}$ is the prehistory of the phase vector at a moment t.

Denote by $X[t_1, t_2]$ the linear normed space of functions $x(t)$ defined on $[t_1, t_2]$ with the range $\mathbb{R}^n$, and the norm $\| x(\cdot) \|_{X[t_1,t_2]}$.

Assume that $f(t, x_t(\cdot))$ maps $t \in [\alpha, \beta]$ and $x_t(\cdot) \in X[\alpha - \tau, t]$ into $\mathbb{R}^n$, and the initial function $\phi \in X[\alpha - \tau, \alpha]$ has properties which guaranty existence of the unique solution $x(\cdot) \in X[\alpha - \tau, \beta]$ of the problem (22.1) - (22.2) . In particular, if $X[t_1, t_2] = C[t_1, t_2]$ is the space of continuous n-dimensional functions with the norm

$$\| x(\cdot) \|_{C[t_1,t_2]} = \max_{t_1 \leq t \leq t_2} \| x(t) \|,$$

then the existence and uniqueness of the solution of the problem (22.1)-(22.2) is guaranted [9, p.55] by the following conditions:

A1: *$f(t, x(\cdot))$ is continuous with respect to t for a fixed $x(\cdot) \in C[\alpha - \tau, \beta]$;*

A2: *$f(t, x(\cdot))$ satisfies the Lipschitz condition with respect to $x(\cdot)$, i.e. there exists L, such that*

$$\| f(t, x^1(\cdot)) - f(t, x^2(\cdot)) \| \leq L \| x^1(\cdot) - x^2(\cdot) \|_{C[\alpha-\tau,t]}$$

for all $t \in [\alpha, \beta]$ and $x^1(\cdot), x^2(\cdot) \in C[\alpha - \tau, \beta]$.

The space $X[t_1, t_2] = \mathbb{R}^n \times Q[t_1, t_2)$, where $Q[t_1, t_2)$ is the space of piece-wise continuous on $[t_1, t_2)$ n-dimensional functions $x(\cdot)$ [18] with the norm $\| x(\cdot) \|_{Q[t_1,t_2)} = \sup_{t_1 \leq t < t_2} \| x(t) \|$ allows us separate finite-dimensional and infinite-dimensional components in prehistory of the phase vector and consider a particular case of the system (22.1)-(22.2):

$$\dot{x} = f(t, x, y), \quad t \in [\alpha, \beta], \tag{22.3}$$

$$x(\alpha) = x^0, \quad y(\alpha) = y^0. \tag{22.4}$$

Here $x = x(t) \in \mathbb{R}^n$, $y = y(t) = \{x_t(s) = x(t + s), \alpha - \tau \leq t + s < t\} \in Q[\alpha - \tau, t)$.

Existence and uniqueness of solutions $x(\cdot) \in Q[\alpha - \tau, \beta)$ of the problem (22.3)-(22.4) are guaranted (see the theorem 14.1) by conditions 0.1 and 0.2:

$$\|f(t, x^1, y^1(\cdot)) - f(t, x^2, y^2(\cdot))\| \leq$$

$$\leq L\|x^1 - x^2\| + M\|y^1(\cdot) - y^2(\cdot)\|_{Q[\alpha-\tau,t)}$$

for all $t \in [\alpha, \beta]$, $x^1, x^2 \in \mathbb{R}^n$, $y^1(\cdot), y^2(\cdot) \in Q[\alpha - \tau, t)$.

Special structure of the system (22.3)-(22.4) allowed us[19] elaborate one-step Runge-Kutta-like and multistep numerical methods similar to the corresponding methods of ODE theory.

---

[18] continuous to the right at discontinuity points

[19] see previous sections

For ODE there are works in which different numerical methods (RK-methods, multistep methods and so on) are described in the framework of a general (unified) numerical schemes. The most general is, apparently, the scheme [20],(see also [10]), which includes all known for author methods.

In this section we extend the scheme [20] on functional differential equations. In this extension the most important component is an intermediate element between the discrete model and FDE: this element is called *interpolation.*

Consider the system (23.1) on the interval $\alpha \leq t \leq \beta$ with the initial condition

$$x(\alpha) = x^0 , \tag{22.5}$$

$$x(\alpha + s) = y^0(s) , \; -\tau \leq s < 0 . \tag{22.6}$$

It is assumed that the system (0.1) satisfies conditions (0.1) and (0.2).

Let a *step size* $\Delta > 0$ (of a discrete model) be such that $\tau/\Delta = m$ is a natural. Denote by $\{\Delta\}$ the set of the discrete model steps.

Consider the equally spaced set of points (grid)

$$\Sigma_\Delta = \{t_i = \alpha + i\Delta \in [\alpha - \tau, \beta] , \; i = -m, \ldots , N\} .$$

Denote

$$\Sigma_\Delta^- = \{t_i \in \Sigma_\Delta, i \leq 0\} , \quad \Sigma_\Delta^+ = \{t_i \in \Sigma_\Delta, i \geq 0\} .$$

The *discrete model* is a discrete function

$$U : t_i \to u(t_i) = u_i \in U_i , \quad i = -m, \ldots , N,$$

where $U_i, i = -m, \ldots , N$ are linear metric spaces with norms $\| \cdot \|_{U_i}$. For example, in most of the described above methods $U_i = R^q$.

The *prehistory* of the discrete model at a moment t_n $(n \geq 0)$ is the set

$$\{u_i\}_{-m}^n = \{u_i \in U_i \; , \quad i = -m, \ldots , n\}.$$

An *interpolational operator* of the discrete model prehistory is a mapping

$$I(\{u_i\}_{-m}^{n}) = v \in V_n \,,$$

where V_n are linear normed spaces with norms $\| \cdot \|_V$.

Definition 22.1 *1) An interpolational operator I satisfies the quasi-Lipschitz condition if there exist $K_1 > 0$ and $\omega_1(\Delta)$ such that for any $\Delta \geq 0$ and all prehistories $\{u_i^{(1)}\}_{-m}^{n}$, $\{u_i^{(2)}\}_{-m}^{n}$ of the discrete model the following inequality holds*

$$\| v^{(1)} - v^{(2)} \|_V \leq K_1 \max_{-m \leq i \leq n} \| u_i^{(1)} - u_i^{(2)} \|_{U_i} + \omega_1(\Delta) \,, \tag{22.7}$$

where $v^{(j)} = I(\{u_i^{(j)}\}_{-m}^{n})$, $j = 1, 2$;

2) the interpolational operator I satisfies the quasi-Lipschitz condition of an order p if there exists $C > 0$ such that $\omega_1(\Delta) \leq C\Delta^p$. □

General linear methods for FDE are based on the *four basic constructions*:

- *An exact value operator,*
- *An advancing model formula,*
- *A starting value function,*
- *An interpolation operator.*

Above we defined the notion of the *interpolational operator*. Let us consider three other elements.

Definition 22.2 *The* exact value operator *is a mapping z such that*

$$z(t_i, \Delta) = z_i \in U_i \,, \quad i = -m, \ldots, N. \tag{22.8}$$

□

It is assumed that the exact value function has a relation[20] with the exact solution $x(t)$ of the initial value

[20] is defined by

problem (23.1),(22.5), (22.6), i.e., there exists a mapping $\Lambda : R^l \to U_i$ such that

$$z_i = \Lambda x_i, \quad i = -m, \ldots, N,$$

where $x_i = x(\alpha + i\Delta)$, $i = -m, \ldots, N$.

Hence, if we know z_i, then to find the values x_i (of the exact solution) one can calculate the inverse operator to Λ. For example, if $U_i = R^q$ and Λ is a linear mapping then $x_i = \Lambda^+ z_i$, $i = -m, \ldots, N$, where Λ^+ is the pseudo-inverse matrix.

Definition 22.3 *The starting value function (or simply the starting function) is a mapping* $U^0 : \Sigma_\Delta^- \times \{\Delta\} \to U_i$ *such that*

$$u_i = \phi(t_i, \Delta), \quad i = -m, \ldots, 0. \tag{22.9}$$

□

Definition 22.4 *The starting value function has an order* p *if there exists a constant* $C > 0$ *such that*

$$\| z_i - \phi(t_i, \Delta) \|_{U_i} \le C\Delta^p, \quad i = -m, \ldots, 0.$$

□

Definition 22.5 *The formula of advancing the model is the algorithm*

$$u_{n+1} = S_n u_n + \Delta\Phi(t_n, I(\{u_i\}_{-m}^n), \Delta), \tag{22.10}$$

where $\Phi : \Sigma_\Delta^+ \times V_n \times \{\Delta\} \to U_{n+1}$ *and* $S_n : U_n \to U_{n+1}$ *are linear operators.* □

Thus a numerical method (the discrete model) is defined by:

- The exact value operator (22.8),
- The formula of advancing the model (22.10),
- The starting value function (22.9),
- The interpolation operator.

Assumption 22.1 *An operator* $\Phi(t_n, v, \Delta)$ *(22.10) satisfies the* quasi-Lipschitz condition *with respect to the second argument, i.e. there exist* $K_2 > 0$ *and* $\omega_2(\Delta)$ *such that for all* $t_n \in \Sigma_\Delta^+$, $\Delta \in \{\Delta\}$, $v^{(1)}, v^{(2)} \in V_n$

$$\| \Phi(t_n, v^{(1)}, \Delta) - \Phi(t_n, v^{(2)}, \Delta) \|_{U_{n+1}} \leq$$
$$\leq K_2 \| v^{(1)} - v^{(2)} \|_{V_n} + \omega_2(\Delta). \qquad (22.11)$$

□

Definition 22.6 *An operator* Φ *satisfies the quasi-Lipschitz condition of an order* p *if there exists a constant* $C > 0$ *such that* $\omega_2(\Delta) \leq C\Delta^p$. □

Remark 22.1 If operators I and Φ satisfy the *quasi-Lipschitz condition* then their superposition[21] $\Phi(t_n, I(\{u_i\}_{-m}^n), \Delta)$ satisfies the *quasi-Lipschitz condition* with respect to the discrete prehistories of the model, because

$$\| \Phi(t_n, I(\{u_i^{(1)}\}_{-m}^n), \Delta) - \Phi(t_n, I(\{u_i^{(2)}\}_{-m}^n), \Delta) \|_{U_{n+1}} \leq$$
$$\leq K_2 \| I(\{u_i^{(1)}\}_{-m}^n) - I(\{u_i^{(2)}\}_{-m}^n) \|_{V_n} + \omega_2(\Delta) \leq$$
$$\leq K_2 K_1 \max_{-m \leq i \leq n} \| u_i^{(1)} - u_i^{(2)} \|_{U_i} + K_2\omega_1(\Delta) + \omega_2(\Delta) \leq$$
$$\leq K \max_{-m \leq i \leq n} \| u_i^{(1)} - u_i^{(2)} \|_{U_i} + \omega(\Delta),$$

where $K = K_2K_1$, $\omega(\Delta) = K_2\omega_1(\Delta) + \omega_2(\Delta)$. □

Definition 22.7 *A method*

1) converges if

$$\delta_n = z_n - u_n \to 0 \quad as \quad \Delta \to 0, \quad n = -m, \ldots, N;$$

2) has the convergence order p *if there exists a constant* $C > 0$ *such that*

$$\| \delta_n \| \leq C\Delta^p, \quad n = -m, \ldots, N.$$

□

[21] the composition

Further we omit sub-indices of norms.

Definition 22.8 *The method (22.10) is stable if the product of operators S^i, $i = k, \ldots, n$, is uniformly bounded for all $n \geq 0$, $0 \leq k \leq n$, i.e. there exists $\tilde{S}$ such that*

$$\| \prod_{i=k}^{n} S^i \| \leq \tilde{S} .$$

□

Definition 22.9 *The* approximation order (residual) *of a method is the discrete function $d_i : \Sigma_\Delta \to U_n$*

$$d_n = z_n - \phi(t_n, \Delta), \;\; n = -m, \ldots, 0,$$

$$d_{n+1} = z_{n+1} - Sz_n - h\Phi(t_n, I(\{z_i\}_{-m}^{n}), \Delta), \;\; n = 0, \ldots, N-1. \tag{22.12}$$

□

Definition 22.10 *The method (22.10) has an* approximation order p *if there exists a constant C such that*

$$\| d_n \| \leq C\Delta^{p+1}, \quad n = 1, \ldots, N.$$

□

The following proposition is valid.

Theorem 22.1 *If*
1) the method (22.10) is stable,
2) the operator Φ satisfies the quasi-Lipschitz condition of an order $p_1 > 0$,
3) the interpolation operator I satisfies the quasi-Lipschitz condition of an order $p_2 > 0$,
4) the starting value function has an order $p_3 > 0$,
5) the method has an approximation order $p_4 > 0$.
Then the method converges and has the convergence order $p = \min\{p_1, p_2, p_3, p_4\}$. □

The proof of the theorem is based on the following proposition.

Lemma 22.1 *If*
1) The method (22.10) is stable,
2) The operator Φ *satisfies the quasi-Lipschitz condition (22.11),*
3) The interpolation operator I *satisfies the quasi-Lipschitz condition (22.7),*
4) $\hat{u}_n$, $n = -m, \dots, N$, *is a solution of the equation*

$$\hat{u}_n = \phi(t_n, \Delta) + r_n, \ \ n = -m, \dots, 0,$$

$$\hat{u}_{n+1} = S\hat{u}_n + \Delta\Phi(t_n, I(\{\hat{u}_i\}_{-m}^n), \Delta) + r_{n+1}, \ n = 0, \dots, N-1, \tag{22.13}$$

where $r_n \in U_n$, $n = -m, \dots, N$, *are disturbances.*
Subsequently the following inequality is fulfilled

$$\|\hat{u}_n - u_n\| \le Ae^{K\tilde{S}(\beta-\alpha)},$$

where $A = C_1 R_0 + \tilde{S}(\beta - \alpha)(\omega(\Delta) + R_1/\Delta)$, $C_1 = \max\{K\tilde{S}(\beta - \alpha), 1\}$, $R_0 = \max\limits_{-m \le i \le 0} \|r_i\|$, $R_1 = \max\limits_{0 \le i \le N} \|r_i\|$.

Proof. Denote $\delta_n = \hat{u}_n - u_n$, $n = -m, \dots, N$, then for $n = 0, \dots, N-1$, we have

$$\delta_{n+1} = \sum_{j=0}^{n} S^{n,j} \Delta \hat{\delta}_j + \sum_{j=0}^{n+1} S^{n+1,j} r_j,$$

where

$$\hat{\delta}_n = \Phi(t_n, I(\{\hat{u}_i\}_{-m}^n), \Delta) - \Phi(t_n, I(\{u_i\}_{-m}^n), \Delta),$$

$$S^{n,k} = \prod_{i=k}^{n} S^i \ \text{ for } \ k \le n,$$

($S^{n,k}$ is the identity operator for $k > n$).

Taking into account remark 22.1 we can estimate

$$\|\hat{\delta}_n\| \le K \max_{-m \le i \le n} \|\delta_i\| + \omega(\Delta).$$

Then from the stability definition it follows that

$$\|\delta_{n+1}\| \le \tilde{S}K\Delta \sum_{j=0}^{n} \max_{-m\le i\le n} \|\delta_i\| + \tilde{S}\Delta \sum_{j=0}^{n} \omega(\Delta) + \tilde{S} \sum_{j=0}^{n+1} r_j ,$$

and

$$\|\delta_{n+1}\| \le \tilde{S}K\Delta \sum_{j=0}^{n} \max_{-m\le i\le n} \|\delta_i\| + \tilde{S}(\beta-\alpha)(\omega(\Delta)+R_1/\Delta) . \tag{22.14}$$

Let us prove, applying the mathematical induction method, the following estimation

$$\|\delta_n\| \le (1+\tilde{S}K\Delta)^n (C_1 R_0 + \tilde{S}(\beta-\alpha)(\omega(\Delta)+R_1/\Delta)) = \\ = A(1+\tilde{S}K\Delta)^n \tag{22.15}$$

for $n = 1, \ldots, N$.

Base of the induction. For $n = 1$ the estimation

$$\|\delta_1\| \le \tilde{S}K\Delta R_0 + \tilde{S}\Delta\omega(\Delta) + \tilde{S}r_1 \le \\ \le \tilde{S}(\beta-\alpha)(KR_0 + \omega(\Delta) + R_1/\Delta) \le \\ \le (1+\tilde{S}k\Delta)(C_1 R_0 + \tilde{S}(\beta-\alpha)(\omega(\Delta)+R_1/\Delta))$$

follows from (22.10) and (22.13).

Induction step. Let the estimation (22.15) be fulfilled for all indices from 1 to n. Let us prove the estimation for $n+1$.

Fix $j \le n$. Let i_0 be an index at which the maximum $\max\limits_{-m\le i\le j} \|\delta_i\|$ is achieved. The following two variants are possible:

a) $i_0 \le 0$ then

$$\max_{-m\le i\le j} \|\delta_i\| = \|\delta_{i_0}\| \le R_0 \le A(1+\tilde{S}K\Delta)^j ;$$

b) $1 \le i_0 \le j$ then

$$\max_{-m\le i\le j} \|\delta_i\| = \|\delta_{i_0}\| \le A(1+\tilde{S}K\Delta)^{i_0} \le A(1+\tilde{S}K\Delta)^j .$$

From these estimations and (22.14) we obtain

$$\|\delta_{n+1}\| \le \tilde{S}K\Delta \sum_{j=0}^{n} A(1+\tilde{S}K\Delta)^j + A = A(1+\tilde{S}K\Delta)^{n+1}.$$

Thus the estimation (22.15) is proved and hence

$$\|\delta_n\| \le A(1+\tilde{S}K\Delta)^n \le Ae^{K\tilde{S}(\beta-\alpha)}.$$

The lemma is proved. □

Proof of Theorem 22.1. Consider in (22.13) that the disturbances r_n equal an approximation error, i.e. $r_n = d_n$. Then, taking into account (22.12), one can obtain the following representation for $\hat{u}_n$

$$\hat{u}_n = \phi(t_n, \Delta) + z_n - \phi(t_n, \Delta) = z_n \quad \text{for} \quad n = -m, \dots, 0,$$

$$\hat{u}_{n+1} = S\hat{u}_n + h\Phi(t_n, I(\{\hat{u}_i\}_{-m}^n), \Delta) - Sz_n -$$
$$- \Delta\Phi(t_n, I(\{z_i\}_{-m}^n), \Delta) \quad \text{for} \; n = 0, \dots, N-1;$$

from which we derive $\hat{u}_{n+1} = z_{n+1}$, $n = 0, \dots, N-1$.

Then from Lemma 22.1 we have the estimation

$$\|z_n - u_n\| \le (C_1 R_0 + \tilde{S}(\beta-\alpha)(\omega(\Delta) + R_1/\Delta))e^{K\tilde{S}(\beta-\alpha)},$$
$$n = 1, \dots, N,$$

which completes the proof of the theorem. □

Remark 22.2 Similar to ODE [10] one can obtain the necessary and sufficient conditions of convergence for an order p in functional differential equations. □

From the described general scheme it follows that numerical methods are distinguished by:

a) Grids and spaces of a discrete models,
b) Interpolational operators,
c) Starting value procedures,
d) Formulas of the model advancing.

Additionally, to calculate an approximation error of a numerical method it is necessary to know an exact value operator.

The core part of the general scheme is the *advancing model formula*, which is often called a *numerical method.*

All known numerical methods for ODE and FDE, can be described in the framework of this general scheme. Consider, for example, how the Stetter method can be included in this scheme.

Example 22.1 Consider FDE of the second order ($x \in \mathbf{R}$)

$$\ddot{x}(t) = f(t, x(t), x_t(\cdot))$$

with the initial conditions

$$x(\alpha) = x_0, \quad \dot{x}(\alpha) = x_1 ,$$

$$x(\alpha + s) = y^0(s) , \; -\tau \leq s < 0 .$$

We can approximate the second order derivative of the solution by the second order difference

$$\frac{u_{n+1} - 2u_n + u_{n-1}}{\Delta^2} = f(t, u_n, u_{t_n}(\cdot)) , \qquad (22.16)$$

Where $u(\cdot) = I(\{u_i\}_{n-m}^n)$ is an interpolation (for example, the piece-wise linear interpolation). From (22.16) we have an explicit 2-step method

$$u_{n+1} = 2u_n - u_{n-1} + \Delta^2 f(t, u_n, u_{t_n}(\cdot)) .$$

Show that the method can be described in the frameworks of the general scheme. Let $\Delta > 0$ is such that $\tau/\Delta = m$ is a natural. Define $t_0 = \alpha + \Delta$ and the partition $t_i = t_0 + i\Delta$, $i = -m, \ldots, N$, ($t_N \leq \beta$). In this case we have 2-dimensional discrete model: $U_n = \mathbf{R}^2$, $u_n = (u_n^{(1)}, u_n^{(2)})$ and

$$u_n^{(2)} = \frac{u_{n+1} - u_n}{\Delta} .$$

Consider an interpolational space consisting of one-dimensional continuous functions $V_n = C[t_n - \tau, t_n + \Delta]$, and introduce an interpolational operator $u^{(1)}(\cdot) = I(\{u_i\}_{n-m}^{n}) = I(\{u_i^{(1)}\}_{n-m}^{n})$ acting only with respect to the first component of the discrete pre-history. Note, if the interpolational operator is piece-wise linear, then it satisfies the Lipschitz condition, and hence, is the quasi-Lipschitz operator.

For the advancing model formula

$$u_{n+1}^{(1)} = u_n^{(1)} + \Delta u_n^{(2)} ,$$

$$u_{n+1}^{(2)} = u_n^{(2)} + \Delta f(t_{n+1}, u_{n+1}^{(1)}, u_{t_{n+1}}(\cdot)) ,$$

the corresponding matrix

$$S_n = \begin{pmatrix} 1 & 0 \\ 0 & 1 \end{pmatrix}$$

has the unit norm, hence the method is stable.

Note, the function

$$\Phi(t_n, I(\{u_i\}_{-m}^{n}), \Delta) = \begin{pmatrix} u_n^{(2)} \\ f(t_n + \Delta, u_n^{(1)} + \Delta u_n^{(2)}, u_{t_{n+1}}(\cdot)) \end{pmatrix}$$

satisfies the Lipschitz condition with respect to the second component.

If an exact solution $x(t)$ is sufficiently smooth, then the method has the second approximation order (for a suitable interpolational operator). If, besides, we can calculate the starting point u_1 with an accuracy of the second order with respect to Δ, then the method has the second order of convergence. □

22.2 Methodology of classification numerical FDE models

One can see that the described above numerical methods for FDE differ from each other by:

a) The grid and the discrete model space;
b) Methods of interpolation of the discrete model;
c) Starting values of the model;
d) Formula of step moving.

And additionally, for introducing the notion of the approximation error necessary the exact value function. Similar to ODE the key moment of this classification, is the formula of step moving the model on one step, which is often called the numerical method.

All known for the author ODE and FDE numerical methods are included into these scheme. Consider several examples of such inclusions.

Explicit Runge-Kutta-like methods

Explicit k-stage Runge-Kutta-like methods for FDE, studied in the section 18.2, are essencially based on separation of finite-dimensional and infinite-dimensional components in the structure of the system (22.3)-(22.4) with conditions 0.1 and 0.2.

The inclusion of these methods into the general scheme can be realized in the following way.

The finite-dimensional discrete model is $U_l = \mathbb{R}^q$, Its dimension coincides with the dimension of the phase vector, i.e. $q = n$. We consider only grids with $t_0 = \alpha$. For $\Delta > 0$ $m = \tau/\Delta$.

The exact values function is given by the phase vector $Z(t_i, \Delta) = z_i = x(t_i)$, $i = -m, ..., N$, that is the mapping Λ is the identity operator.

Take the starting values of the model coinciding with the values $u_l = x(t_l)$, $l = -m, ..., 0$. Therefore the initial approximation error is $d_l = 0, l = -m, ..., 0$.

The specific feature of the Runge-Kutta-like methods is that to define at a moment t_l the formula for moving the model on one step necessary to make both the interpolation of the discrete model prehistory on the interval $[t_l - \tau, t_l]$, and the extrapolation on $[t_l, t_l + a\Delta]$, $a > 0$. Let

$a > 0$ be fixed, in the character of the interpolation space take $V = Q[t_l - \tau, t_l + a\Delta]$. In the section 18.1 we described methods of constructing the interpolation operator (in case of Runge-Kutta-like methods it consists of the interpolation and extrapolation). The simplest of them is the piecewise constant interpolation–extrapolation: $I(\{u_i\}^l_{-m}) = I(\{u_i\}^l_{l-m}) = v(\cdot) \in Q[t_l - \tau, t_l + a\Delta]$,

$$v(t) = \begin{cases} u_i, \ t \in [t_i, t_{i+1}), \ i = l - m, \dots, l - 1, \\ u_l, \ t \in [t_i, t_l + a\Delta). \end{cases}$$

This operator satisfies the Lipschitz condition with the constant $K_1 = 1$.

Further we assume that I is the *consistent* operator, i.e. $v(t_i) = u_i$ for all $i = l - m, \dots, l$, $v(\cdot) = I(\{u_i\}^l_{-m})$.

The formula for one-step advancement is

$$u_{l+1} = u_l + \Delta \sum_{i=1}^{k} \sigma_i \Phi_i(t_l, I(\{u_i\}^l_{-m}), \Delta), \tag{22.17}$$

where

$$\Phi_1(t_l, I(\{u_i\}^l_{-m}), \Delta) = f(t_l, u_l, v_{t_l}(\cdot)), \tag{22.18}$$

$$\Phi_i(t_l, I(\{u_i\}^l_{-m}), \Delta) =$$

$$= f(t_l + a_i\Delta, u_l + \Delta \sum_{j=1}^{i-1} b_{ij} \Phi_j(t_l, I(\{u_i\}^l_{-m}), \Delta), v_{t_l + a_i h}(\cdot)), \tag{22.19}$$

σ_i, a_i, b_{ij} $(i = 1, \dots, k, \ j = 1, \dots, i - 1)$ — are parameters of the method, $|a_i| \leq a, \ i = 1, \dots, k$.

Lemma 22.2 *The function*

$$\Phi(t_l, I(\{u_i\}^l_{-m}), \Delta) = \sum_{i=1}^{k} \sigma_i \Phi_i(t_l, I(\{u_i\}^l_{-m}), \Delta)$$

satisfies the Lipschitz condition with respect to the second variable.

Proof. Denote $v^1 = I(\{u_i^1\}_{-m}^l)$, $v^2 = I(\{u_i^2\}_{-m}^l)$. We prove that applying the induction method by i that the function $\Phi_i(t_l, v, \Delta)$ satisfies the Lipschitz condition with respect to the second variable with the constant L_i.

The basis. Because the mapping f satisfies the Lipschitz condition and I is the consistent operator, therefore

$$\|\Phi_1(t_l, v^2, \Delta) - \Phi_1(t_l, v^1, \Delta)\| =$$

$$= \|f(t_l, u_l^2, v_{t_l}^2(\cdot)) - f(t_l, u_l^1, v_{t_l}^1(\cdot))\| \leq$$

$$\leq L\|u_l^2 - u_l^1\| + M\|v^2 - v^1\|_{V_l} \leq (L + M)\|v^2 - v^1\|_{V_l},$$

That is $L_1 = L + M$.

The induction step. Due to the Lipschitz condition and the induction assumption we have

$$\|\Phi_i(t_l, v^2, \Delta) - \Phi_i(t_l, v^1, \Delta)\| =$$

$$= \|f(t_l + a_i\Delta, u_l^2 + \Delta \sum_{j=1}^{i-1} b_{ij}\Phi_j(t_l, v^2(\cdot), \Delta),$$

$$v_{t_l+a_i\Delta}^2(\cdot)) - f(t_l + a_i\Delta, u_l^1 + \Delta \sum_{j=1}^{i-1} b_{ij}\Phi_j(t_l, v^1(\cdot), \Delta), v_{t_l+a_i\Delta}^1(\cdot))\| \leq$$

$$\leq L\Delta \sum_{j=1}^{i-1} |b_{ij}| \|\Phi_j(t_l, v^2(\cdot), \Delta) - \Phi_j(t_l, v^1(\cdot), \Delta)\| + M\|v^2 - v^1\|_{V_l} \leq$$

$$\leq L\Delta \sum_{j=1}^{i-1} |b_{ij}| L_j \|v^2 - v^1\|_{V_l} + M\|v^2 - v^1\|_{V_l}.$$

Thus

$$L_i = (L\Delta \sum_{j=1}^{i-1} |b_{ij}| L_j + M).$$

From the Lipschitz condition for the function $\Phi_i(t_l, v, \Delta)$ follows the Lipschitz condition for the function $\Phi(t_l, v, \Delta)$ with respect to the second variable with the constant $L_\Phi = \sum_{i=1}^{k} |\sigma_i| L_i$. □

Stability conditions for ERK-like methods are satisfied, because S_l is the identical operator.

Therefore, it follows from theorem 22.1 that the Runge-Kutta-like method has the convergence order p, if its local approximation order is $p+1$.

The local approximation order of the residual in the general scheme is defined by the function d_l in (22.12), i.e. it depends on the one-step advancement formula and the method of interpolation.

In section 18.2, we studied the explicit Runge-Kutta-like method and due to the coincidence of dimensions of the model and the phase vector the residual is defined in another way than in (22.12):

Definition 22.11 *The residual (without interpolation) of the explicit Runge-Kutta-like method is*

$$\tilde{d}_{l+1} = x_{l+1} - x_l - \Delta\Phi(t_l, x(\cdot), \Delta), \ \ l = 0, \dots, N-1,$$

where

$$\Phi(t_l, x(\cdot), \Delta) = \sum_{i=1}^{k} \sigma_i \Phi_i(t_l, x(\cdot), \Delta),$$

$$\Phi_1(t_l, x(\cdot), \Delta) = f(t_l, x_l, x_{t_l}(\cdot)),$$

$$\Phi_i(t_l, x(\cdot), \Delta) = f(t_l + a_i\Delta, x_l + \Delta\sum_{j=1}^{i-1} b_{ij}\Phi_j(t_l, x(\cdot), \Delta), x_{t_l+a_i\Delta}(\cdot)).$$

The explicit Runge-Kutta-like method has a *local residual order (without interpolation)* $p+1$, if there exists a constant C, such that

$$\| \tilde{d}_l \| \leq C\Delta^{p+1} \text{ for all } l = 1, \dots, N. \qquad (22.20)$$

The local order of the residual (without interpolation) is defined by the set of parameters $\sigma_i, a_i, b_{ij} (i = \overline{1,k(1)}, j = 1, i-1(1))$, according to the same rules as for ODE. This can be verified with the application of methods of the i-smooth analysis.

Definition 22.12 *Interpolation operator I has the order p on a class of functions X, containing the exact solution, if there exist a constant C, such that for all $x(\cdot) \in X$, $l = 0, \dots, N$ the following inequality is valid*

$$\|I\{x_i\}_{l-m}^{l} - x(\cdot)\|_{V_l} \leq C\Delta^p \,. \qquad (22.21)$$

For example, the operator interpolating by degenerate splines of degree $p-1$ has p-th order of interpolation on the class of p-times continuous differentiable functions.

Lemma 22.3 *If the ERK-like method has the residual (without interpolation) of the order $p+1$ and interpolation operator has the order p on the class of functions, containing the exact solution, then the method converges and has the local approximation order $p+1$ in the sense of the definition (22.21).*

Proof. According to the residual definition (22.12), for explicit Runge-Kutta-like methods for all $l = 0, \dots, N-1$

$$\|d_{l+1}\| = \|x_{l+1} - x_l - \Delta\Phi(t_l, I(\{x_i\}_{-m}^{l}), \Delta)\| \leq$$

$$\leq \|x_{l+1} - x_l - \Delta\Phi(t_l, x(\cdot), \Delta)\| +$$

$$+ \|\Delta\Phi(t_l, I(\{x_i\}_{-m}^{l}), \Delta) - \Delta\Phi(t_l, x(\cdot), \Delta)\|,$$

then from (22.20), lemma 22.2 and (22.21) we have

$$\|d_{l+1}\| \leq C\Delta^{p+1} \,.$$

□

Continuous one-step methods

In papers [6, 36] for numerically solving the problem (22.1)–(22.2) methods were elaborated approximating the exact solution not only at points of the grid, but also at all intermediate points. Consider similar constructions for functional differential equations. Fix a grid $t_l = \alpha + l\Delta$, $l = 0, 1, ..., N$, and $\Delta = (\beta - \alpha)/N > 0$.

Definition 22.13 *A continuous one-step method is the model*

$$u(t) = \phi(t), \quad t \in [\alpha - \tau, \alpha], \tag{22.22}$$

$$u(t_l + r\Delta) = u(t_l) + r\Delta\Psi(t_l, u_{t_l}(\cdot), r), \quad 0 \le r \le 1, \tag{22.23}$$

where $\Psi(t, u_t(\cdot), r)$ *is the operator which maps* $t \in [\alpha, \beta]$, $u_t(\cdot) \in C[\alpha - \tau, t]$ *and* $r \in [0, 1]$ *into a vector* $\mathbb{R}^n$.

In the paper [36]to prove convergence of method (22.22)–(22.23) with the order p require that the mapping $\Psi(t, u_t(\cdot), r)$ would be continuous with respect to r, Lipschitz with respect to $u_t(\cdot)$, and guarantees the residual order p:

$$\|\frac{x(t_l + r\Delta) - x(t_l)}{\Delta} - \Psi(t_l, x_{t_l}(\cdot), r)\| \le C\Delta^p, \tag{22.24}$$

for all $r \in [0, 1]$, $l = 0, 1, ..., N - 1$.

Consider how methods (22.22)–(22.23) can be included into the general scheme.

Introduce the spaces $U_l = C[\alpha - \tau, t_l]$ and models — sets $u_l = u_{t_l}(\cdot) \in U_l$. Interpolational spaces for such models are not necessary, and we put $V_l = U_l$.

Define an operator S_l: $U_l \to U_{l+1}$ by the formula

$$S_l u(\cdot) = \hat{u}(t) = \begin{cases} u(t), & t < t_l, \\ u(t_l), & t \ge t_l. \end{cases}$$

S_l is the linear and $\|S_l\| = 1$.

The exact value function is defined by the exact solution of the problem (22.1)–(22.2):

$$z_l = x_{t_l}(\cdot) \in U_l = C[\alpha - \tau, t_l].$$

The formula of one-step advancement has the form

$$u_{l+1} = S_l u_l + \Delta\Phi(t_l, u_l, \Delta), \tag{22.25}$$

where

$$\Phi(t_l, u_l, \Delta) =$$

$$= \tilde{u}(t) = \begin{cases} 0, & t < t_l, \\ (t - t_l)\Psi(t_l, u_{t_l}(\cdot), \frac{t-t_l}{\Delta}), & t_l \le t \le t_{l+1}. \end{cases}$$

Taking into account the forms of u_l, S_l Φ one can see that the one-step advancement formula is the equivalent form of presentation of the method (22.23). $\|S_l\| = 1$ therefore the method is stable.

If Ψ satisfies the Lipschitz condition with respect to the second variable, then Φ also satisfies the Lipschitz condition with respect to the second variable. If Ψ has the approximation order p in the sense (22.24), then the approximation order of the method (22.25) is also equal to p, and it follows from the theorem 22.1 that the method converges with the order p.

Continuous methods till the fourth convergence order are constructed in [36].

Multistep and other methods

Let $k \ge 1$ be a natural. Take $t_0 = \alpha + (k-1)\Delta$, where $\Delta > 0$. Let at the moment $t_l = t_0 + l\Delta$ the approximate value of the solution of the system (22.3)–(22.4) is defined by $u_l^1 = u^1(t_l) \in \mathbb{R}^n$. Let on the set $\{u_i^1\}_{l-m}^l$, $m = \tau/\Delta$ an interpolation operator $I(\{u_i^1\}_{l-m}^l)$ with the range $Q[t_l - \tau, t_l]$ is given.

Explicit k-step method (see section 19.1) is the algorithm

$$u_l^1 = \sum_{j=1}^{k} a_j u_{l-j}^1 + \Delta \sum_{j=1}^{k} b_j f(t_{l-j}, u_{l-j}^1, I(\{u_i^1\}_{l-j-m}^{l-j})). \tag{22.26}$$

By analogy with the ODE case multistep methods can be reduced to one-step methods by increasing the model dimension.

Consider the vector u_l of the space $U_l = U = \mathbb{R}^q$, $q = kn$:

$$u_l = (u_l^1, u_{l-1}^1, ..., u_{l-k+1}^1)^t = (u_l^1, u_l^2, ..., u_l^k)^t,$$

and let $u(t) \in Q[t_l - \tau - k\Delta, t_l]$ be the result of acting the interpolation operator $I(\{u_i\}_{l-m}^{l}) = I(\{u_i^1\}_{l-m-k}^{l})$.

Then the method (22.26) can be presented in the form

$$u_{l+1} = Su_l + \Delta\Phi(t_l, I(\{u_i\}_{-m}^{l}), \Delta), \tag{22.27}$$

where

$$S = \begin{pmatrix} a_1 & a_2 & \dots & a_{k-1} & a_k \\ 1 & 0 & \dots & 0 & 0 \\ . & . & \dots & . & . \\ 0 & 0 & \dots & 1 & 0 \end{pmatrix},$$

$$\Phi(t_l, I(\{u_i\}_{-m}^{l}), \Delta) = \begin{pmatrix} \Phi^1 \\ 0 \\ . \\ 0 \end{pmatrix};$$

$$\Phi^1 = \sum_{j=1}^{k} b_j f(t_{l-j+1}, u_l^j, u_{t_{l-j+1}}(\cdot)).$$

The eigenvalues of the matrix S are roots of the characteristic equation

$$\lambda^k + a_1\lambda^{k-1} + \dots + a_k = 0,$$

so the method is stable if and only if when all roots of this equation have absolute value no greater than 1 and there are no multiple roots with the absolute value equal to 1.

22.3 Necessary and sufficient conditions of convergence with order p

In this section we assume that the space of discrete models is the finite-dimensional: $U_l = \mathbb{R}^q$ for all $l = -m, \dots, N$, and let step-by-step formula (22.10) linear operators S_l be given by the matrix $S \in \mathbb{R}^{q \times q}$, i.e.

$$u_{l+1} = Su_l + \Delta\Phi(t_l, I(\{u_i\}_{-m}^{l}), \Delta). \tag{22.28}$$

All the above mentioned methods, except the continuous one, satisfy these conditions. If the residual d_l tends to the zero as Δ tends to zero, for all $l = 1, ..., N$, then the equality $Z(t, 0) = SZ(t, 0)$ is valid, and the unit is the eigenvalue of the matrix S. Assume that this condition is satisfied. If, besides, the method is stable, then the matrix S can not have eigenvalues outside the unite circle, and eigenvalues with the absolute value can not originate in a Jordan series. Denote eigenvalues with the unit absolute value as: $\lambda_1 = 1,\ \lambda_2, ..., \lambda_k$. Then the canonic Jordan form of S will have the block-diagonal structure

$$S = Tdiag\left\{\begin{pmatrix} 1 & 0 & ... & 0 \\ 0 & 1 & ... & 0 \\ . & . & ... & . \\ 0 & 0 & ... & 1 \end{pmatrix}, \begin{pmatrix} \lambda_2 & 0 & ... & 0 \\ 0 & \lambda_2 & ... & 0 \\ . & . & ... & . \\ 0 & 0 & ... & \lambda_2 \end{pmatrix}, ..., \right.$$

$$\left. \begin{pmatrix} \lambda_k & 0 & ... & 0 \\ 0 & \lambda_k & ... & 0 \\ . & . & ... & . \\ 0 & 0 & ... & \lambda_k \end{pmatrix}, Y \right\} T^{-1},$$

T is a non-singular matrix. Thus

$$S = E_1 + \lambda_2 E_2 + ... + \lambda_k E_k + \tilde{E},$$

where

$$E_1 = Tdiag\{E, 0, 0, ..., 0\}T^{-1},$$
$$E_2 = Tdiag\{0, E, 0, ..., 0\}T^{-1},$$
$$...$$
$$\tilde{E} = Tdiag\{0, 0, 0, ..., Y\}T^{-1},$$

E is the unit matrix.

Definition 22.14 *The method (22.28) is consistent with the order p, if there is a constant C such that the residual satisfies the relations*

$$||d_l|| \le C\Delta^p \text{ for } l = 0, ..., N,$$

$$||E_1(d_0 + d_1 + ... + d_l) + d_{l+1}|| \le C\Delta^p \text{ for } l = 0, ..., N-1.$$

If the method has the residual order p, then it is consistent with the order p. The reverse is not true, examples of such methods see in [20].

Definition 22.15 *The method (22.28) is strongly stable, if all eigenvalues of the matrix S have absolute values less than 1, except, may be, the simple root $\lambda_1 = 1$.*

Theorem 22.2 *Let the method (22.28) be strongly stable, the function Φ satisfies the quasi-Lipschitz condition with the order $p > 0$, the interpolation operator I satisfies the quasi-Lipschitz condition with the order p, the starting values have the order p. Then the method converges with the order p if and only if it is consistent with the order p.*

Proof. Let the method converge with the order p. Taking into account the relation $E_1 S = E_1$ we have

$$E_1(d_0 + ... + d_l) + d_{l+1} = \sum_{j=0}^{l+1} S^{l+1-j} d_j - (S - E_1) \sum_{j=0}^{l} S^{l-j} d_j \,,$$

from which for $l \leq N - 1$

$$\|E_1(d_0 + ... + d_l) + d_{l+1}\| = (1 + \|S - E_1\|)\|D\|_S,$$

where $\|S\|$ is the subordinated norm of the matrix,

$$\|D\|_S = \max_{0 \leq l \leq N} \|\sum_{j=0}^{l} S^{l-j} d_j\|.$$

Take in (22.13) disturbances equal to the residual, i.e. $r_l = d_l$, then from the estimation of the lemma 17.1 obtain the consistence with the order p.

In reverse, let the method be consistent with the order p. The method is strongly stable, hence $S = E_1 + \tilde{E}$ and for $0 \leq j \leq l - 1$ we have

$$S^{l-j} = E_1 + \tilde{E}^{l-j}. \tag{22.29}$$

Then for $r_l = d_l$

$$||R||_S = \max_{0 \le l \le N} ||\sum_{j=0}^{l} S^{l-j} d_j|| =$$

$$= \max_{0 \le l \le N} ||\sum_{j=0}^{l-1} (E_1 d_j + \tilde{E}^{l-j} d_j) + d_l|| \le$$

$$\le \max_{0 \le l \le N-1} ||\sum_{j=0}^{l} E_1 d_j + d_{l+1}|| + \frac{1}{1 - ||\tilde{E}||} \max_{0 \le l \le N-1} ||d_l||.$$

From the consistency property and the condition $||\tilde{E}|| < 1$, we have $||R||_S = O(\Delta^p)$, which together with estimation of the lemma 17.1 gives convergence of the method with the order p. □

Similar to ODE case, the condition of the strong stability in the theorem 22.2 can be relaxed till the condition of stability, if the residual has a smoothness property.

Definition 22.16 *The residual has the smoothness property of the order p, if the residual can be expanded into the corresponding Taylor series with respect to Δ:*

$$d_{l+1} = d^0(t_l) + d^1(t_l)\Delta + ... + d^p(t_l)\Delta^p + O(\Delta^p),\ l = 0, ..., N-1,$$

where $d^j(t)$ are $(p-j+1)$-times continuous differentiable functions.

The smoothness property of the residual depends on the smoothness of the exact value function and the functional Φ in the one-step advancement formula of the model, which, in one's turn, depend on smoothness of the functional in the right-hand side of FDE and its solution. For the system (22.3) - (22.4) these conditions can be obtained with application of the technique of i-smooth analysis.

Lemma 22.4 *If the residual has a smoothness property of the order p, then the method is consistent with the order p*

if and only if

$$\|d_l\| \leq C\Delta^p \text{ for } l = 0, ..., N, \quad \text{and } E_1 d^p(t) = 0 \text{ for } t \in (\alpha, \beta]. \tag{22.30}$$

Proof. From (22.30) and the smoothness property of the order p we obtain

$$d_j = d^p(t_{j-1})\Delta^p + O(\Delta^p), \ E_1 d_j = O(\Delta^p), \ j = 1, ..., N,$$

from this result follows the consistent condition with the order p.

Reverse, let the method be consistent with the order p. It means that $E_1(d_1 + ... + d_l) = O(\Delta^p)$. Take $t \in (\alpha, \beta]$ and consider a grid with the step Δ and a natural l, such that $t - t_0 = l\Delta$. Then

$$E_1(d_1 + ... + d_l) = \Delta^p E_1 \sum_{j=1}^{l} d^p(t_{j-1}) + O(\Delta^p) =$$

$$= \Delta^{p-1} E_1 \int_{t_0}^{t} d^p(s)ds + O(\Delta^p).$$

Then we obtain

$$E_1 \int_{t_0}^{t} d^p(s)ds = 0,$$

differentiating this equality we have $E_1 d^p(t) = 0$. □

Theorem 22.3 *If the method (22.28) is stable, the function Φ satisfies the quasi-Lipschitz condition with the order $p > 0$, the interpolation operator I satisfies the quasi-Lipschitz condition with the order $p > 0$, starting values have the order p, residual has smoothness property with the order, then the method converges with the order p, if and only if it is coordinated with the order p.*

Proof. Comparing to the proof of theorem 22.2, some changes should be made in the second part. The equality (22.29) is changed by

$$S^{l-j} = E_1 + \lambda_2^{l-j} E_2 + ... + \lambda_k^{l-j} E_k + \tilde{E}^{l-j}.$$

Then by taking into account the lemma 22.4, obtain

$$\|R\|_S = \max_{0 \le l \le N} \|\sum_{j=0}^{l} S^{l-j} d_j\| \le$$

$$\le \max_{0 \le l \le N} \|\Delta^p E_2 \sum_{j=1}^{l} \lambda_2^{l-j} d^p(t_{j-1}) + ... +$$

$$+\Delta^p E_k \sum_{j=1}^{l} \lambda_k^{l-j} d^p(t_{j-1}) + \sum_{j=0}^{l} \tilde{E}^{l-j} d_j\| + O(\Delta^p).$$

From the equality

$$\sum_{j=1}^{l} \lambda^{l-j} d(t_{j-1}) = \frac{1-\lambda^l}{1-\lambda} d(t_0) + \sum_{j=1}^{l} \frac{1-\lambda^{l-j}}{1-\lambda} (d(t_j) - d(t_{j-1}))$$

we have that the expression $\sum_{j=1}^{l} \lambda^{l-j} d(t_{j-1})$ is uniformly bounded, if $|\lambda| = 1$, $\lambda \ne 1$ and $d(t)$ has a bounded variation.

Then, taking into account that $\|\tilde{E}\| < 1$, we have $\|R\|_S = O(\Delta^p)$. This together with the estimation of the lemma 17.1 gives convergence of the method with the order p. □

22.4 Asymptotic expansion of the global error

In this section we study the asymptotic expansion of the global error with respect to the value of step discretion (see [10, 11, 20])

For ordinary differential equations the principal term of the expansion of the global error satisfies a linear system of differential equations, which is the adjoint system to the original system. For functional differential equations the principal term of the expansion of the global error also satisfies a linear system of functional differential equations. The asymptotic expansion of the global error is the theoretical basis for constructing a wide class of interpolational methods [10] and *for realizing the automatic step size control procedure.*

In the first part of the section we consider one-step methods, in the second part the results are generalized on arbitrary strongly stable methods.

One-step method

In the subsection we consider a one-step method for the problem (22.1)–(22.2)

$$u_l = \phi(t_l), \quad l = -m, \dots, 0,$$

$$u_{l+1} = u_l + \Delta\Phi(t_l, u(\cdot), \Delta), \quad l = 0, \dots, N-1, \quad (22.31)$$

with the grid

$$t_0 = \alpha,\ t_l = t_0 + l\Delta,\ l = 0, \dots, N,\ \Delta = \tau/m,$$

And the interpolation operator $u(\cdot) = I(\{u_i\}_{-m}^{l})$ which maps $\mathbb{R}^{n(l+m)}$ into the spaces $X[\alpha - \tau, t_l + a\Delta]$, $a > 0$.

The one-step method is the particular case of the general scheme of the section 22.1, in which spaces of the discrete model coincide with the phase space: $U_i = \mathbb{R}^n$, $l = -m, \dots, N$; the exact value function is given by the phase vector $Z(t_i, \Delta) = z_i = x(t_i)$, $i = -m, ..., N$; the starting values coincide with the exact values $u_l = x(t_l)$, $l = -m, ..., 0$; the operator S in the step-formula is the identity matrix: $S = E$.

For example, it can be an explicit or implicit Runge-Kutta-like method for the system (22.3)

$$\Phi(t_l, u(\cdot), \Delta) = \sum_{i=1}^{k} \sigma_i \, h_i,$$

$$h_i = h_i(t_l, u(\cdot)) = f(t_l + a_i\Delta, u_l + \Delta \sum_{j=1}^{k} b_{ij}h_j, u_{t_l+a_i\Delta}(\cdot)).$$

Make additional assumptions defining the interpolation operator I and the one-step advancement function Φ for $\Delta \to 0$.

Fix t_* and decrease the step Δ, increasing l such that $t_l = t_0 + l\Delta = t_*$.

Let $x(t)$ be a function of the set $X[t_0 - \tau, t_*]$.

Condition 22.1 *Consistency of the limit of the prehistory interpolation operator I.*

$$I(\{x(t_i)\}_{-m}^{l}) \to x(t) \text{ as } \Delta \to 0 \text{ in the space } X[t_0 - \tau, t_*].$$

Condition 22.2 *Linearity of the prehistory interpolation operator I.*

For all prehistories $\{u_i^1\}_{-m}^{l}$ and $\{u_i^2\}_{-m}^{l}$ and constants c^1 and c^2 the following condition is fulfilled

$$I(\{c^1u_i^1 + c^2u_i^2\}_{-m}^{l}) = c^1I(\{u_i^1\}_{-m}^{l}) + c^2I(\{u_i^2\}_{-m}^{l}).$$

Condition 22.3 *Boundedness of the prehistory interpolation operator I.*

There is a constant $C = \|I\|$, such that for all prehistories $\{u_i\}_{-m}^{l}$ the following condition holds

$$\|I(\{u_i\}_{-m}^{l})\| \leq C\|\{u_i^1\}_{-m}^{l}\|.$$

Operator I realizes the interpolation by degenerate splines with extrapolation by continuation then for the operator these conditions are satisfied. From boundedness follows the Lipschitz condition of the operator I and hence the quasi-Lipschits condition with arbitrary order p.

Denote (if the limit exists)

$$\Phi(t_*, u(\cdot), 0) = \lim_{\Delta \to 0} \Phi(t_*, u(\cdot), \Delta).$$

Condition 22.4 *Consistency of the step advancement function.*

For the exact solution of the system (22.1) the following condition is held

$$\Phi(t_*, x_{t_*}(\cdot), 0) = f(t_*, x_{t_*}(\cdot)).$$

This condition is satisfied for a Runge-Kutta-like method, if the method has the residual order $p > 0$.

Condition 22.5 *The functional $\Phi(t_*, u(\cdot), \Delta)$ is a Frechet differentiable with respect to the second variable, moreover, the Frechet derivative $\Phi'_{u(\cdot)}(t_*, u(\cdot), \Delta)$ is continuous at zero with respect to Δ.*

The Frechet differentiability of the functional Φ in Runge-Kutta-like methods follows from the Frechet differentiability of the functional f. From the Frechet differentiability, it follows that the functional Φ satisfies the Lipschits condition with respect to the second variable.

Theorem 22.4 *If an approximation error of a method (the residual) can be presented in the form*

$$d_l = \gamma(t)\Delta^{p+1} + o(\Delta^{p+1}), \quad l = 0, \ldots, N-1, \qquad (22.32)$$

where $\gamma(t)$ is a continuous on $[t_0, t_0+\theta]$ function. Then under the conditions 22.1–22.5 the global error of the method can be presented in the form

$$u_l - x_l = e(t_l)\Delta^p + o(\Delta^p), \qquad (22.33)$$

where the function $e(t)$ satisfies on $[t_0, t_0 + \theta]$ the equation

$$\dot{e}(t) =< G(t), e_t(\cdot) > -\gamma(t) \qquad (22.34)$$

with the initial conditions

$$e(t) = 0, \ t \leq t_0.$$

Here $G(t) = \Phi'_{x(\cdot)}(t, x(\cdot), 0)$ *is the Frechet derivative of the functional* $\Phi(t, x(\cdot), 0)$ *in space* $X[t_0 - \tau, t]$, $< G(t), e_t(\cdot) >$ *is the result of acting* $G(t)$ *on* $e_t(\cdot) \in X[t_0 - \tau, t]$.

Proof. Consider the value

$$\hat{u}_l = u_l - e(t_l)\Delta^l, \ l = 0, ..., N,$$

as the result of acting new numerical method. Then for $l \geq 0$ obtain

$$\hat{u}_{l+1} = u_{l+1} - e(t_{l+1})\Delta^l =$$

$$= \hat{u}_l + [e(t_l) - e(t_{l+1})]\Delta^p + \Delta\Phi(t_l, I(\{\hat{u}_i + e(t_i)\Delta^p\}^l_{-m}), \Delta) =$$

$$= \hat{u}_l + \Delta\hat{\Phi},$$

where

$$\hat{\Phi} = \Phi(t_l, \hat{I}(\{\hat{u}_i\}^l_{-m}), \Delta) + [e(t_l) - e(t_{l+1})]\Delta^{p-1},$$

$$\hat{I}(\{\hat{u}_i\}^l_{-m}) = I(\{\hat{u}_i + e(t_i)\Delta^p\}^l_{-m}).$$

Calculate the residual of this new method

$$\hat{d}_l = x_{l+1} - x_l - \Delta\Phi(t_l, \hat{I}(\{x_i\}^l_{-m}), \Delta) - [e(t_l) - e(t_{l+1})]\Delta^p =$$

$$= \gamma(t_l)\Delta^{p+1} + \Delta[\Phi(t_l, I(\{x_i\}^l_{-m}), \Delta) -$$

$$-\Phi(t_l, I(\{x_i + e(t_i)\Delta^p\}^l_{-m}), \Delta)] + [e(t_{l+1}) - e(t_l)]\Delta^p + o(\Delta^{p+1}) =$$

$$= \gamma(t_l)\Delta^{p+1} - \Delta < \Psi'_{x(\cdot)}(t_l, I(\{x_i\}^l_{-m}), \Delta), I(\{e(t_i)\Delta^p\}^l_{-m}) > +$$

$$+e'(t_l)\Delta^{p+1} + o(\Delta^{p+1}) =$$

$$= \Delta^p[e'(t_l) - < \Psi'_{x(\cdot)}(t_l, x_{t_l}(\cdot), 0), e_{t_l}(\cdot) > +\gamma(t_l)] + o(\Delta^p).$$

If the function $e(t)$ satisfies the equation (22.34), then the residual will have the order higher than p (in the sense of definition of section 18.1). Operator I satisfies the Lipschits condition, because the starting values have zero error. The function Φ is Lipschits with respect to the second variable, hence according to theorem 22.1, the new method has a convergence order greater than p. □

General scheme. Strongly stable finite-dimensional methods

Consider the variant of the general scheme with the finite-dimensional space of models: $U_l = \mathbb{R}^q$, $S_l = S \in \mathbb{R}^{q\times q}$. This is the case discussed in the section 22.3.

Similar to one-step methods consider additional assumptions on the interpolation operator I and the step advancement function Φ when $\Delta \to 0$.

Fix t_* and decrease the step Δ increasing l such, that $t_l = t_0 + l\Delta = t_*$.

Let $e(t)$ be q-dimension vector-function [22] of the linear normed space $X^q[t_0 - \tau, t_*]$. For every l, such that $t_l = t_0 + l\Delta = t_*$ the interpolation operator defines the elements $v_l = I(e\{t_i\}_{-m}^{l})$ of the interpolation space V_l. Let there be a mapping Π from the space V_l into a linear normed space V_{t_*}. Define the value of the interpolation operator on $e(t)$ by the relation

$$I(e(\cdot)) = \lim_{l\to\infty} \Pi I(e\{t_i\}_{-m}^{l}) \tag{22.35}$$

We assume that the following conditions are valid

Condition 22.6 *The limit in (22.35) exists for the exact value function $z(t) = \Lambda x(t)$ for $t \in [\alpha, \beta]$.*

Condition 22.7 *The operator of prehistory interpolation and its supplementation are linear.*

Similarly, define the limit value of the step advancement function $\Phi(t_l, v_l, \Delta)$ when Δ tends to zero such that $t_l = t_0 + l\Delta = t_*$.

[22] defined on the interval $[t_0 - \tau, t_*]$

Let there be a mapping Π of the space V_l into a linear normed space V_{t_*} such that for all sequences $v_l \in V_l$ there exist $\lim_{l\to\infty} \Pi v_l = v^* \in V_{t_*}$. Define $\Phi(t_*, v^*, 0)$ by the relation

$$\Phi(t_*, v^*, 0) = \lim_{l\to\infty} \Phi(t_l, v_l, \Delta). \tag{22.36}$$

In some cases we use the following conditions

Condition 22.8 *The limit in (22.36) exists.*

Condition 22.9 *The functional $\Phi(t_*, v, \Delta)$ is the Frechet differentiable with respect to the second variable, and moreover, the Frechet derivatives $\Phi'_v(t_*, v, \Delta)$ are continuous at the zero with respect to Δ.*

Theorem 22.5 *Let the approximation error of the method (residual) be presented in the form*

$$d_l = \gamma(t_l)\Delta^p + o(\Delta^p), \quad l = -l, \ldots, 0, \tag{22.37}$$

$$d_l = d_p(t_l)\Delta^p + d_{p+1}(t_l)\Delta^{p+1} + o(\Delta^{p+1}), \quad l = 0, \ldots, N, \tag{22.38}$$

where $\gamma(t)$ is a continuous on $[t_0 - \tau, t_0]$ function, $d_p(t)$ is a continuous differentiable on $[t_0, \beta]$ function, and $d_{p+1}(t)$ is continuous on $[t_0, \beta]$ function. Additionally, if the method is strongly stable, I is the quasi-Lpschits operator with the order greater than p, Φ is the quasi-Lipschits function with respect to the second variable with the order greater than p, and

$$E_1 d_p(t) = 0, \; t \in [t_0, \beta],$$

(the last condition follows from the consistency condition with the order p). Then under the conditions 22.6–22.9 the global error can be presented in the form

$$u_l - x_l = (e(t_l) + \varepsilon_l)\Delta^p + o(\Delta^p), \tag{22.39}$$

and the function $e(t)$ satisfies on the interval $[t_0, \beta]$ the linear functional differential equation

$$\dot{e}(t) = E_1 < G(t), I(e_t(\cdot)) > -E_1 d_{p+1}(t) - (E - S - E_1)^{-1}\dot{d}_p(t) \tag{22.40}$$

with the initial conditions

$$e(t_0) = -\gamma(t_0) - \varepsilon_0, \tag{22.41}$$

$$e(t) = -\gamma(t), \ t < t_0. \tag{22.42}$$

Here $G(t) = \Phi'_v(t, I(z_t(\cdot)), 0)$ *is the Frechet derivative of the functional* $\Phi(t, I(z_t(\cdot)), 0)$ *with respect to the second variable; the values* ε_l, *tend to the zero as* $\to \infty$, *and are defined by the relations*

$$\varepsilon_0 = (E_1 - E)\gamma(t_0) + (E - S + E_1)^{-1} d_p(t_0), \tag{22.43}$$

$$\varepsilon_l = (S - E_1)^l \varepsilon_0. \tag{22.44}$$

Proof. Consider the value

$$\hat{u}_l = u_l - e(t_l)\Delta^l, \ l < 0,$$

$$\hat{u}_l = u_l - (e(t_l) + \varepsilon_l)\Delta^l, \ l = 0, ..., N,$$

as the result of acting a new numerical method.

Then for $t < t_0$ the residual of this method $\hat{d}_l = z_l - \hat{u}_l = (\gamma(t_l) + e(t_l))\Delta^p + o(\Delta^p)$ has the order greater than p, if (22.42)holds.

For $l = 0$ we have

$$z_0 - \hat{u}_0 = z_0 - u_0 + (e(t_0) + \varepsilon_0)\Delta^p =$$

$$= (\gamma(t_0) + e(t_0) + \varepsilon_0)\Delta^p + o(\Delta^p). \tag{22.45}$$

For $l \geq 0$, according to the step-by-step formula we have

$$\hat{u}_{l+1} = u_{l+1} - (e(t_{l+1} + \varepsilon_{l+1})\Delta^l =$$

$$= \hat{u}_l + [Se(t_l) + S\varepsilon_l - e(t_{l+1}) - \varepsilon_{l+1}]\Delta^p +$$

$$+\Delta\Phi(t_l, I(\{\hat{u}_i + (e(t_i) + \varepsilon_i)\Delta^p\}_{-m}^l), \Delta) = S\hat{u}_l + \Delta\hat{\Phi},$$

where

$$\hat{\Phi} = \Phi(t_l, \hat{I}(\{\hat{u}_i\}_{-m}^l), \Delta) + [Se(t_l) + S\varepsilon_l - e(t_{l+1}) - \varepsilon_{l+1}]\Delta^{p-1},$$

$$\hat{I}(\{\hat{u}_i\}_{-m}^{l}) = I(\{\hat{u}_i + (e(t_i) + \varepsilon_{-m}^{l})\Delta^p\}_l).$$

Calculate the residual of the new method for $l \geq 0$

$$\hat{d}_l = z_{l+1} - Sz_l - \Delta\Phi(t_l, \hat{I}(\{z_i\}_{-m}^{l}), \Delta) -$$

$$-[Se(t_l) + S\varepsilon_l - e(t_{l+1}) - \varepsilon_{l+1}]\Delta^p = d_p(t_l)\Delta^p + d_{p+1}\Delta^{p+1} +$$

$$+\Delta\Phi(t_l, I(\{z_i\}_{-m}^{l}), \Delta) - \Delta\Phi(t_l, I(\{z_i + (e(t_i) + \varepsilon_i)\Delta^p\}_{-m}^{l}), \Delta) +$$

$$+[e(t_{l+1}) + \varepsilon_{l+1} - Se(t_l) - S\varepsilon_l]\Delta^p + o(\Delta^{p+1}). \quad (22.46)$$

From (22.45)–(22.46) it follows that the residual is consistent with the order greater than p, if the same time the following conditions are held

$$\gamma(t_0) + e(t_0) + \varepsilon_0 = 0, \quad (22.47)$$

$$d_p(t_l) + e(t_l) - Se(t_l) + \varepsilon_{l+1} - S\varepsilon_l = 0, \quad (22.48)$$

$$E_1[\dot{e}(t_l) - < G(t), I(e_{t_l}(\cdot)) > + d_{p+1}(t_l)] = 0, \quad (22.49)$$

$$E_1\varepsilon_l = 0. \quad (22.50)$$

Solve this system of equations. Condition (22.48) is satisfied, if at the same time the following conditions are held

$$\varepsilon_{l+1} = S\varepsilon_l, \quad (22.51)$$

$$d_p(t) + e(t) - Se(t) = 0 \text{ for } t \geq t_0. \quad (22.52)$$

The condition (22.52) for $t = t_0$ has the form

$$d_p(t_0) = -e(t_0) + Se(t_0). \quad (22.53)$$

write (22.52) in the equivalent form:

$$d_p(t) = (S - E)e(t) \text{ for } t \geq t_0, \quad (22.54)$$

where E is the identity matrix.

Prove the relation

$$(E - S + E_1)^{-1}(E - S) = E - E_1. \quad (22.55)$$

From the definition of strong stability follows that $S = E_1 + \tilde{E}$ (see section 22.3). Hence $SE_1 = E_1^2 = E_1$. The obvious identity is valid

$$(E - S + E_1)(E - E_1) = E - S,$$

Hence, taking into account non-singularity of the matrix $E - S - E_1$ [23] have (22.55).

From (22.54) and (22.55) follows

$$(E - S + E_1)^{-1} d_p(t) = (E_1 - E)e(t). \qquad (22.56)$$

Multiplying (22.47) by $(E - E_1)$ obtain, taking into account (22.56),

$$(E-E_1)\varepsilon_0 = (E_1-E)\gamma(t_0)+(E-S+E_1)^{-1}d_p(t_0). \quad (22.57)$$

Chose the value ε_0 from relation (22.43). Then from the consistency condition with the order p follows $E_1\gamma(t_0) = 0$ and $E_1 d_p(t_0) = 0$, therefore

$$E_1\varepsilon_0 = 0, \qquad (22.58)$$

and the relation (22.57), and therefore also the relation (22.47) is hold.

Chose ε_l from the relation (22.51) (condition (22.50) is automatically satisfied). Then from the condition (22.58) follows

$$\varepsilon_l = S^l\varepsilon_0 = (E_1 + S - E_1)^l\varepsilon_0 =$$
$$= (E_1 - E)\gamma(t_0) + (E - S + E_1)^{-1}d_p(t_0),$$

and from the strong stability it also follows that values ε_l, tend to the zero as $l \to \infty$.

Take $e(t)$ as the solution of the equation(22.49) with the initial conditions (22.41) – (22.42).

Consider $v(t) = E_1 e(t)$, then because of (22.56),

$$e(t) = v(t) - (E - S + E_1)^{-1}d_p(t).$$

Differentiating this relation obtain, due to (22.49), equation (22.40). □

[23] the nonsingularity follows from the fact that $S - E_1$ cannot have characteristic values equal to the zero

Remark 22.3 The variant of the method with weakly stable matrix S is considered in similar way [11].

Remark 22.4 By analogy with [10, 11] one can derive equations for next terms of the global error of the interpolation operator and coefficients of the residual expansion, under more strong assumptions on smoothness of the functional Φ and some additional assumptions on starting values.

23 Algorithms with variable step-size and some aspects of computer realization of numerical models

Software tools for numerical solving ODEs realize procedures with automatic step size control. Specific modifications of the algorithms, described in two previous sections, allow us develop step size control algorithms for FDE.

As the example of such modification we describe algorithms and prove the corresponding convergence theorem for explicit Runge-Kutta-like methods with interpolation by degenerate splines and extrapolation by continuation. We also consider error estimations for algorithms, when functionals of the right-hand sides of FDE systems are calculated approximately: it should be taken into account, for example, for functionals containing integrals of the prehistory.

These algorithms along with the automatic step size procedure (with the help of estimation of the principal term of local error of embedded methods) are fundament for elaborating FDE software.

23.1 ERK-like methods with variable step

Similar to the begining of the chapter 2 consider the FDE system

$$\dot{x}(t) = f(t, x(t), x_t(\cdot)), \tag{23.1}$$

With the initial condition

$$x(t_0) = x_0, \tag{23.2}$$

$$x_{t_0}(\cdot) = \{y^0(s), -\tau \le s < 0\}, \tag{23.3}$$

with the standard assumptions on the functional f.

Set on $[t_0, t_0 + \theta]$ a non-uniform *grid* $t_{l+1} = t_l + \Delta_l$, $l = 0, 1, \ldots, N-1$, where $\Delta_l > 0$ is the *step*. Denote $\Delta_{max} = \max\limits_l \Delta_l$, $\Delta_{min} = \min\limits_l \Delta_l$. Assume that $\Delta_{max} \le K\Delta_{min}$, K is the fixed number.

Introduce the *discrete numerical model* of the system (23.1), denoting by $u_l \in \mathbb{R}^\ell$ the approximation of the exact solution $x(t_l) = x_l$ at the point t_l.

The specific feature of the system (23.1) is that at the moment t_l the adequate discrete model should take into account the *prehistory of the discrete model*, i.e. the set $\{u_i\}_l = \{u_i, t_i \in [t_l - \tau, t_l]\}$. For methods with variable step size necessary take into account an enlarged prehistory.

Definition 23.1 *Extended prehistory of the discrete model is the set* $\{u_i\}_l^+ = \{u_i,\ t_i \in [t_l - 2\tau, t_l]\}$.

Definition 23.2 *On the set of extended discrete model prehistories* $\{u_i\}_l^+$ *an interpolation–extrapolation operator* IE *is given, if for* $a > 0$

$$IE(\{u_i\}_l^+) = u(\cdot) \in Q[t_l - \tau, t_l + a\Delta_{max}). \tag{23.4}$$

Further we assume that

1) IE is the *consistent* operator, i.e.

$$u(t_i) = u_i,\ t_i \in [t_l - \tau, t_l]; \tag{23.5}$$

2) The operator IE satisfies the Lipschitz condition, i.e. there exists a constant L_{IE} such that for any two sets of discrete prehistories $\{u_i^1\}_l^+$ and $\{u_i^2\}_l^+$ and for all $t \in [t_l - \tau, t_l + a\Delta_{max})$ the following inequality is satisfied

$$\|u^1(t) - u^2(t)\| \leq L_{IE} \max_{t_l - 2\tau \leq t_i \leq t_l} \|u_i^1 - u_i^2\|, \quad (23.6)$$

For $u^1(\cdot) = IE(\{u_i^1\}_l^+)$, $u^2(\cdot) = IE(\{u_i^2\}_l^+)$.

Definition 23.3 *An interpolation–extrapolation operator IE has the order p on a set of functions $X \subseteq Q[t_l - \tau, t_l + a\Delta_{max})$, if there exists a constant C_{IE}, such that for all $x(\cdot) \in X$ and $t \in [t_l - \tau, t_l + a\Delta_{max})$*

$$\|x(t) - \tilde{x}(t)\| \leq C_{IE}\Delta^p, \quad (23.7)$$

where $\tilde{x}(\cdot) = IE(\{x_i\}_l^+)$, $x_i = x(t_i)$, $t_i \in [t_l - 2\tau, t_l]$.

The operator of piece-wise constant interpolation and extrapolation $IE : \{u_i\}_l \to u(\cdot)$:

$$u(t) = \begin{cases} u_i, \ t \in [t_i, t_{i+1}), \\ u_l, \ t \in [t_l, t_l + a\Delta_{max}) \end{cases}$$

is the consistent operator, and satisfies the Lipschits condition with $L_{IE} = 1$. Moreover, it has the first order on the set of Lipschits functions $x(t)$. Examples of higher order interpolation–extrapolation operators are discussed in the section 23.2.

Definition 23.4 *Explicit k-stage Runge-Kutta-like method (ERK) with variable step is the discrete model of the form*

$$u_0 = x_0, \quad u_{l+1} = u_l + \Delta_l \Psi(t_l, u(\cdot)), \quad l = 1, \ldots, N-1, \quad (23.8)$$

$$u(\cdot) = IE(\{u_i\}_l^+), \quad \Psi(t_l, u(\cdot)) = \sum_{i=1}^{k} \sigma_i h_i, \quad (23.9)$$

$$h_1 = h_1(t_l, u(\cdot)) = f(t_l, u_l, u_{t_l}(\cdot)), \quad (23.10)$$

$$h_i = h_i(t_l, u(\cdot)) = f(t_l + a_i\Delta_l, u_l + \Delta_l \sum_{j=1}^{i-1} b_{ij}h_j, u_{t_l+a_i\Delta_l}(\cdot)),$$
$$|a_i| \le a,\ i = 1, \dots, k. \tag{23.11}$$

The Butcher table of the method is

$$\begin{array}{c|cccc} 0 & 0 & 0 & \dots & 0 \\ a_2 & b_{21} & 0 & \dots & 0 \\ \cdot & \cdot & \cdot & \dots & \cdot \\ a_k & b_{k1} & b_{k2} & \dots & 0 \\ \hline & \sigma_1 & \sigma_2 & \dots & \sigma_k \end{array}$$

Denote $a = \max\{|a_i|, 1 \le i \le k\}$, $\sigma = \max\{|\sigma_i|\}$, $b = \max\{|b_{ij}|\}$.

The following proposition is valid

Lemma 23.1 *Functionals $h_1, h_2, ..., h_k$, and Ψ satisfy the Lipschits condition with respect to $u(\cdot)$.*

The proof of this lemma is similar to the proof of the corresponding proposition (lemma 18.2) of the section 18.2.

Corollary 23.1 *If the interpolation–extrapolation operator IE satisfies the Lipschits condition with a constant L_{IE}, then the function $\Phi(\{u_i\}_l^+) = \Psi(t_l, IE(\{u_i\}_l^+))$ satisfies the Lipschits condition with respect to the discrete model prehistory, i.e. there exists the constant $L_\Phi = L_\Psi L_{IE}$, such that*

$$\|\Phi(\{u_i^1\}_l^+) - \Phi(\{u_i^2\}_l^+)\| \le L_\Phi \max_{t_l-2\tau \le t_i \le t_l} \|u_i^1 - u_i^2\|.$$

Let $x_l = x(t_l)$, $l = 0, 1, \dots, N$ be the values of the exact solution $x(t)$ of the problem (23.1)–(23.3) at points t_l.

The *residual* of the ERK method on the exact solution $x(\cdot) = \{x(t), t \in [t_0 - \tau, t_0 + \theta]\}$ at points $t_l, l = 0, 1, \dots, N$ are the functionals

$$g_l(x(\cdot)) = \frac{x_{l+1} - x_l}{\Delta_l} - \Psi(t_l, x(\cdot)), \tag{23.12}$$

where

$$\Psi(t_l, x(\cdot)) = \sum_{i=1}^{i-1} \sigma_i h_i,$$

$$h_i = f(t_l + a_i\Delta_l, x_l + \Delta_l \sum_{j=1}^{i-1} b_{ij}h_j, x_{t_l+a_i\Delta_l}(\cdot)).$$

For ODE the residual order is usually defined on some class of problems, for example on the class of sufficiently smooth problems. For FDE the problem (23.1) – (23.3) is defined by the triple $(f, x_0, y^0(\cdot))$, i.e. the functional of the right-hand side of system (23.1); the finite-dimensional x_0 and the infinite-dimensional $y^0(\cdot)$ components of the initial data. Let $F = \{(f, x_0, y^0(\cdot))\}$ be a subset (class of problems) of such triples, which satisfy the conditions of the section 11.1 and let $X(F)$ be the corresponding set of solutions $x(t)$ of the problems (23.1) – (23.3).

We say that *ERK-method has the approximation order (of the residual) p on the class of problems F*, if for any $x(\cdot) \in X(F)$ there exists a constant $C_{x(\cdot)}$ such that

$$\|g_l(x(\cdot))\| \leq C_{x(\cdot)}\Delta_{max}^p, \quad l = 0, \ldots, N-1. \qquad (23.13)$$

We say that ERK-method (23.8) – (23.11) for the problem (23.1) – (23.3) *converges*, if

$$\max_{1\leq l\leq N} \|u_l - x(t_l)\| \to 0 \text{ as } \Delta_{max} \to 0,$$

and the method *converges with the order p*, if there exists a constant C_u, such that

$$\|x_l - u_l\| \leq C_u\Delta_{max}^p, \quad l = 0, 1, \ldots, N. \qquad (23.14)$$

For ODE the convergence order of ERK-methods is defined only by the approximation order. For FDE it also depends on the order of the interpolation–extrapolation operator. The following proposition is valid

Theorem 23.1 *If the ERK-method (23.8) – (23.11) has the approximation order* $p_1 > 0$ *and the interpolation–extrapolation operator* IE *has the order* $p_2 > 0$. *Then the ERK-method converges and has the convergence order* $p = \min\{p_1, p_2\}$.

Proof. Denote $\delta_l = u_l - x_l$, then from (23.8) follows that $\delta_0 = 0$.

For $l = 0, \ldots, N-1$ we have

$$\delta_{l+1} = \delta_l + \Delta_l[\Psi(t_l, IE(\{u_i\}_l^+)) - \Psi(t_l, IE(\{x_i\}_l^+))] +$$

$$+[x_l - x_{l+1} + \Delta_l \Psi(t_l, x(\cdot))] + \Delta_l[\Psi(t_l, I(\{x_i\}_l^+)) - \Psi(t_l, x(\cdot))].$$

Applying to the function $\Phi(\{u_i\}_l^+) = \Psi(t_l, IE(\{u_i\}_l^+))$ the corollary 23.1, and taking into account (23.13), (23.11), obtain

$$||\delta_{l+1}|| \leq ||\delta_l|| + \Delta_l L_\Phi \max_{t_l - 2\tau \leq t_i \leq t_l} ||\delta_i|| + C_{x(\cdot)} \Delta_l^{p+1} + L_\Psi C_{IE} \Delta_l^{p+1}. \tag{23.15}$$

Let us apply the mathematical induction method with respect to l to prove the estimation

$$||\delta_l|| \leq (1 + \Delta_{max}(L_\Phi + 1))^l (C_{x(\cdot)} + L_\Psi C_{IE}) \Delta_{max}^p, \tag{23.16}$$

$$l = 0, \ldots, N.$$

Basis. For $\delta_0 = 0$ the estimation is valid.

Induction step. Assuming that the estimation (23.16) is valid for all $\leq l$. Prove it for $l + 1$.

From (23.15) and the inductive assumption we have

$$||\delta_{l+1}|| \leq (1 + \Delta_{max}(L_\Phi + 1))^l (C_{x(\cdot)} + L_\Psi C_{IE}) \Delta_{max}^p +$$

$$+\Delta_{max} L_\Phi (1 + \Delta_{max}(L_\Phi + 1))^{l_0} (C_{x(\cdot)} + L_\Psi C_{IE}) \Delta_{max}^p +$$
$$+(C_{x(\cdot)} + L_\Psi C_{IE}) \Delta_{max}^{p+1},$$

where $l_0 \leq l$ is the index at which the maximum $\max\limits_{t_l - 2\tau \leq t_i \leq t_l} ||\delta_i||$ is achived. Then

$$||\delta_{l+1}|| \leq (1 + \Delta_{max}(L_\Phi + 1))^l (1 + \Delta_{max} L_\Phi + \Delta_{max}) \cdot$$

$$\cdot (C_{x(\cdot)} + L_\Psi C_{IE})\Delta_{max}^p,$$

i.e the estimation (23.16) is valid for $l+1$.

Because the number of variable steps is the finite number

$$N \leq \frac{\theta}{\Delta_{min}} \leq \frac{K\theta}{\Delta_{max}},$$

Therefore from (23.16) we obtain the estimation

$$\|x_l - u_l\| \leq (1 + \Delta_{max}(L_\Phi + 1))^N (C_{x(\cdot)} + L_\Psi C_{IE})\Delta_{max}^p \leq$$

$$\leq (C_{x(\cdot)} + L_\Psi C_{IE}) e^{K(L_\Phi + 1)\theta} \Delta_{max}^p, \quad l = 0, 1, \ldots, N,$$

which completes the proof of the theorem. □

The key moment of constructing high order methods (along with the properties of the interpolation-extrapolation operator) is the effective defining the residual order and hence, choice of the Butcher matrix coefficients.

All results of the section 18.3 can be expanded on the variant of variable step.

23.2 Methods of interpolation and extrapolation of discrete model prehistory

Similar to the section 23.1. we consider the interval $[t_0, t_0 + \theta]$ a grid. Remember that the *interpolation operator I of the extended discrete model prehistory* is the mapping $I : \{u_i\}_l^+ \to u(\cdot) \in Q[t_l - \tau, t_l]$.

The operator I has the approximation error p on the exact solution if there exist constant C_1, C_2 such that for all $l = 0, 1, ..., N$ and $t \in [t_l - \tau, t_l]$ the following inequality is valid

$$\|x(t) - u(t)\| \leq C_1 \max_{i \geq 0, t_l - 2\tau \leq t_i \leq t_l} \|u_i - x_i\| + C_2 \Delta_{max}^p. \tag{23.17}$$

The example of the second order interpolation operator on a sufficiently smooth exact solution is the piece-wise linear interpolation:

$$I : \{u_i\}_l^+ \to u(t) =$$

$$= \begin{cases} ((t-t_i)u_{i+1} + (t_{i+1}-t)u_i)\frac{1}{\Delta_i}, t \in [t_i, t_{i+1}],\ t_{i+1} \in [t_l - \tau, t_l], \\ y^0(t_0 - t),\ t \in [t_l - \tau, t_0). \end{cases} \tag{23.18}$$

Note, if $t \in [t_i, t_{i+1}]$, $t_{i+1} \in [t_l - \tau, t_l]$, but $t_i \notin [t_l - \tau, t_l]$, then the extended prehistory is the essensial part in construction of $u(t)$.

Consider the most general method of interpolation by functions, piece-wise composed by polynomials of degree p (p is a natural). Assume that $\tau/\Delta_{max} \geq p$, i.e. the maximal step has a relatively small value.

Divide the interval $[\max\{t_0, t_l - \tau\}, t_l]$ from the right on subintervals $[t_{l_{i+1}}, t_{l_i}]$, $i = 0, 1, ..., k-1$, unifying $p+1$ neighbourhood points of the grid: $t_{l_0} = t_l$, $t_{l_1} = t_{l-p}, ..., t_{l-ip}, ...$. The last subinterval $[t_{l_k}, t_{l_{k-1}}]$ can contain less than $p+1$ points t_i of the grid[24] so add to it points t_i of the interval $[t_l - 2\tau, t_l - \tau)$. If points are not sufficient because of $t_i < t_0$, then construct fictive points $t_i = t_0 + i\Delta_0$, $i < 0$, values of the midel at these points take as $u_i = y^0(t_i - t_0)$.

On every interval $[t_{l_{i+1}}, t_{l_i}]$ construct interpolation polynomial $L_p(t) = L_p^i(t)$[25]:

$$L_p^i(t) = \sum_{n=0}^{p} u_{l_i - n} \prod_{j=l_i-p;\ j \neq l_i - n}^{l_i} \frac{t - t_j}{t_{l_i - n} - t_j}.$$

Define the interpolation operator of the model prehistory

$$I : \{u_i\}_l \to u(t) = \begin{cases} L_p^i(t),\ t_{l_{i+1}} \leq t < t_{l_i},\ i = 0, ..., l-1, \\ y^0(t_0 - t),\ t \in [t_l - \tau, t_0). \end{cases} \tag{23.19}$$

Theorem 23.2 *Let the exact solution $x(t)$ be $(p+1)$-times continuous differentiable on the interval $[t_0 - \tau, t_0 + \theta]$, then the operator of interpolation by degenerate splines of p-th degree has the interpolation error of $(p+1)$-th order.*

---

[24] of the interval $[\min\{t_0, t_l - \tau\}, t_l]$,

[25] by the data $u_{l_i - p}, u_{l_i - p + 1}, ..., u_{l_i}$

Proof. Fix arbitrary $l \in 1, \ldots, N$ and $t \in [\max\{t_0, t_l - \tau\}, t_l]$. If t belongs to subinterval $[t_{l_{i+1}}, t_{l_i}]$ then

$$\|x(t) - u(t)\| = \|x(t) - L_p^i(t)\| \leq$$

$$\leq \|x(t) - \hat{L}_p^i(t)\| + \|L_p^i(t) - \hat{L}_p^i(t)\|,$$

where $\hat{L}_p^i(t)$ is the interpolation polynomial of p-th degree, constructed according to the values x_j on the interval $[t_{l_{i+1}}, t_{l_i}]$:

$$\hat{L}_p^i(t) = \sum_{n=0}^{p} x_{l_i-n} \prod_{j=l_{i+1};j\neq l_i-n}^{l_i} \frac{t - t_j}{t_{l_i-n} - t_j}.$$

The interpolation error of the exact solution can be calculated by the formula [31, p. 133]

$$\|R_p(t)\| = \|x(t) - \hat{L}_p^i(t)\| = \|\frac{x^{(p+1)}(\xi)}{(p+1)!} \prod_{j=0}^{p} (t - t_{l_i-j})\|,$$

with $\xi \in [t_{l_{i+1}}, t_{l_i}]$. Thus if $t \in [t_{l_{i+1}}, t_{l_i}]$, then

$$\|\prod_{j=0}^{p} (t - t_{l_i-j})\| \leq \Delta_{max} \cdot \Delta_{max} \cdot (2\Delta_{max}) \cdots (p\Delta_{max}) = \Delta_{max}^{p+1} p!$$

and

$$\|R_p(t)\| \leq \frac{M_{p+1}}{(p+1)!} p! \, \Delta_{max}^{p+1} = \frac{M_{p+1}}{p+1} \Delta_{max}^{p+1}, \qquad (23.20)$$

where $M_{p+1} = \max_{t_0-\tau \leq t \leq t_0+\theta} \|x^{(p+1)}(t)\|$. Besides we have estimate

$$\|L_p^i(t) - \hat{L}_p^i(t)\| = \sum_{n=0}^{p} \|u_{l_i-n} - x_{l_i-n}\| \prod_{j=l_i-p;j\neq l_i-n}^{l_i} \frac{|t - t_j|}{|t_{l_i-n} - t_j|}$$

$$\leq \max_{l_i-p \leq n \leq l_i} \|u_n - x_n\| \, p! \frac{\Delta_{max}}{\Delta_{min}}. \qquad (23.21)$$

From(23.20) and (23.21) we obtain

$$\|x(t) - L_p^i(t)\| \leq C_1 \max_{l_i - p \leq n \leq l_i} \|u_n - x_n\| + C_2 \Delta^{p+1}. \quad (23.22)$$

Here $C_1 = Kp!$, $C_2 = \dfrac{M_{p+1}}{p+1}$. □

Besides the interpolation (23.19) one can apply:

– the interpolation by the Hermite polynomials with multiple points, using the condition $\dot{u}_l = f(t_l, u_l, u_{t_l}(\cdot))$;

– other methods of interpolating discrete model prehistory.

IRK-methods require to know at a moment t_l necessary the prehistory of the model $u_{t_l+a}(\cdot)$ for $a > 0$, i.e. necessary to make an extrapolation on the interval $[t_l, t_l + a]$.

Remember that for $a > 0$ an *extrapolation* operator E of the model prehistory is a mapping $E : \{u_i\}_l^+ \to u(\cdot) \in Q[t_l, t_l + a\Delta_l]$.

The extrapolation of the model prehistory has the approximation order p on the exact solution if there exist constants C_3, C_4 such that for all $a > 0$, and for all $l = 0, 1, ..., N-1$ and $t \in [t_l, t_l + a\Delta_l]$ the following inequality holds

$$\|x(t) - u(t)\| \leq C_3 \max_{t_l - 2\tau \leq t_i \leq t_l} \|u_i - x_i\| + C_4 (\Delta_{max})^p. \quad (23.23)$$

One of the methods of defining an extrapolation operator is the extrapolation by *continuation* of interpolation polynomial:

$$E : \{u_i\}_l \to u(t) = L_p^0(t), \quad t \in [t_l, t_l + a\Delta_l], \qquad (23.24)$$

where $L_p^0(t)$ is the interpolation polynomial of p-th dgree, constructed according to the values u_j on the interval $[t_{l-p}, t_l]$:

$$L_p^0(t) = \sum_{n=0}^{p} u_{l-n} \prod_{j=l-p; j \neq l-n}^{l} \frac{t - t_j}{t_{l-n} - t_j}.$$

Theorem 23.3 *Let the exact solution $x(t)$ be $(p+1)$-times continuous differentiable on the interval $[t_0 - \tau, t_0 + \theta]$, then the operator of extrapolation by continuating the interpolation polynomial of p-th degree has the approximation error of the order $p+1$.*

Proof. Fix $t \in [t_l, t_l + a\Delta]$. Then

$$\|x(t) - u(t)\| = \|x(t) - L_p^0(t)\| \leq$$

$$\leq \|x(t) - \hat{L}_p^0(t)\| + \|L_p^0(t) - \hat{L}_p^0(t)\|,$$

where $\hat{L}_p^0(t)$ is the interpolation polynomial of p-th degree, constructed according to the values x_j on the interval $[t_{l-p}, t_l]$.

Then

$$\|x(t) - \hat{L}_p^o(t)\| \leq \frac{M_{p+1}}{(p+1)!} \prod_{j=0}^{p} (t - t_{l-j}) \leq$$

$$\leq \frac{M_{p+1}}{(p+1)!} \prod_{j=0}^{p} (a\Delta_{max} + j\Delta_{max}) \leq C_4(\Delta)_{max}^{p+1},$$

where

$$C_4 = \frac{M_{p+1}}{(p+1)!} \prod_{j=0}^{p} (a+j).$$

Moreover,

$$\|L_p^0(t) - \hat{L}_p^0(t)\| \leq$$

$$\leq \max_{l-p \leq n \leq l} \|u_n - x_n\| sum_{n=0}^{p} \prod_{j=0; j \neq n}^{p} \frac{(a\Delta_{max} + j\Delta_{max})}{\Delta_{min}} \leq$$

$$\leq C_3 \max_{l-p \leq n \leq l} \|u_n - x_n\|.$$

Here

$$C_3 = (p+1) \prod_{j=0}^{p} (a+j).$$

The theorem is proved. □

Unifying statements of theorems 23.1, 23.2 and taking into account the definition of the interpolation-extrapolation operator IE obtain

Corollary 23.2 *The operator of interpolation by degenerate splines of p-th degree extrapolated by the continuating has the order $p+1$ on the sufficiently smooth solution.*

23.3 Choice of the step size

Consider for FDE numerical methods the *Runge rule of practical estimating the error* .

Assume that applying a method of order p we obtain, realizing passage from the zero step to the first step, the following asymptotic representation of the error

$$\varepsilon_1 = x_1 - u_1 = C\Delta^{p+1} + O(\Delta^{p+2}).$$

Make two steps of the length Δ and one step of the length 2Δ. Error for two steps consists of transition error of the first step and the local error of the second step [26]:

$$\varepsilon_2 = x_2 - u_2 = x_2 - \hat{u}_2 + \hat{u}_2 - u_2,$$

where $\hat{u}_2$ is the numerical solution obtained at the moment t_2 from the initial positions $\{t_1, x(t_1), x_{t_1}(\cdot)\}$. Then

$$\hat{u}_2 - u_2 = x_1 - u_1 + \Delta(\Psi(IE(\{x_i\}_1)) - \Psi(IE(\{u_i\}_1))).$$

From the lemma 23.1 and its corollary follows the validity of the inequality

$$\|\Psi(IE(\{x_i\}_1)) - \Psi(IE(\{u_i\}_1))\| \leq L_\Phi |x_1 - u_1|,$$

hence

$$\varepsilon_2 = C\Delta^{p+1} + C\Delta^{p+1}(1 + O(\Delta)) + O(\Delta^{p+2}) =$$

[26] so it can be expressed as

$$= 2C\Delta^{p+1} + O(\Delta^{p+2}).$$

For the value w, obtained by applying the same method with the step 2Δ, we have

$$x_2 - w = C(2\Delta)^{p+1} + O(\Delta^{p+2}).$$

Subtracting one expression from another we obtain

$$u_2 - w \approx 2C\Delta^{p+1}(2^p - 1),$$

hence

$$\varepsilon_2 \approx \frac{u_2 - w}{2^p - 1}.$$

This formula gives not only estimation of the approximation error, but also the possibility of obtaining more accurate result:

$$\hat{u}_2 = u_2 + \frac{u_2 - w}{2^p - 1}.$$

This procedure is called the *Richardson* extrapolation. It is the basis for a class of ODE *extrapolation methods*, among which the most powerful is the Greg-Bulirsh-Shtern algorithm (GBS)[10].

On the basis of the approximate estimation of the error

$$\varepsilon \approx \frac{||u_2 - w||}{2^p - 1}$$

one can organize the procedure of automatic step size control, which guarantees the requiring admissible accuracy *tol*. Denote the approximation by *err* [27].

If the initial value of step is fixed Δ_{old}, then the algorithm realizes calculation of two steps of this length and one step of double length. Calculate error

$$err = \frac{1}{2^p - 1} \max_{i=1,\ldots,n} \frac{|u_{2,i} - w_i|}{d_i},$$

where index i denotes the coordinate of the vector, d_i is the scale factor. For $d_i = 1$ we obtain the absolute error,

[27] as it is used in standard software

for $d_i = |u_{2,i}|$, obtain the relative error. Often other norms and scale factors are applied.

From the relations

$$err = C(2\Delta_{old})^{p+1},$$

$$tol = C(2\Delta_{new})^{p+1}$$

obtain the formula of the new step

$$\Delta_{new} = \left(\frac{tol}{err}\right)^{\frac{1}{p+1}} \Delta_{old}.$$

There are possible two variants:

1) $\Delta_{new} < \Delta_{old}$. Then the previous step is not acceptable and is decreasing till the value Δ_{new}.

2) $\Delta_{new} > \Delta_{old}$. Then two previous steps are acceptable and the hext step will be new, possibly greater.

For practical realization of this procedure avoid the abrupt change (increasing or decreasing) of the step, i.e. the formula for a new step is much complicated:

$$\Delta_{new} = \min\left\{facmax, \max\left\{facmin, fac\left(\frac{tol}{err}\right)^{\frac{1}{p+1}}\right\}\right\}\Delta_{old},$$

where the value fac is usually equal to 0.8, and values $facmax$ and $facmin$ define maximal and minimal coefficients of increasing and decreasing the step.

Besides, chosing the step necessary to satisfy the condition

$$\Delta_{min} \leq \Delta_{new} \leq \Delta_{max}.$$

The idea of *embedded Runge-Kutta formulas* is based on the following methodology: for estimation of the error and realizing the automatic step size control one can use not only two different steps by one and the same method, but one step with respect to two methods of different orders. High effectiveness is achieved if coefficients a_i, b_{ij} of the Butcher matrix, and concequently h_i, of these two methods coincide (such methods are called the *embedded methods*). Method of p-th order

$$u_{l+1} = u_l + \Delta_l \sum_{i=1}^{k} \sigma_i h_i(t_l, u(\cdot))$$

is the basic method, and the method of $p+1$-th order

$$\hat{u}_{l+1} = u_l + \Delta_l \sum_{i=1}^{k} \hat{\sigma}_i h_i(t_l, u(\cdot))$$

is used for the error estimation.

The simple example of such methods is the pair — the improved Euler method and the Runge-Kutta-like method of the third order:

$$\begin{array}{c|ccc}
0 & & & \\
1 & 1 & & \\
\frac{1}{2} & \frac{1}{4} & \frac{1}{4} & \\
\hline
u_{l+1} & \frac{1}{2} & \frac{1}{2} & 0 \\
\hline
\hat{u}_{l+1} & \frac{1}{6} & \frac{1}{6} & \frac{4}{6}
\end{array}$$

This method is called *Runge-Kutta-Fehlberg method of the second–third order* and is denoted as RKF 2(3).

Another more accurate, six-stage *Runge-Kutta-Fehlberg of fourth–fifth order* method RKF 4(5):

$$\begin{array}{c|cccccc}
0 & & & & & & \\
\frac{1}{4} & \frac{1}{4} & & & & & \\
\frac{3}{8} & \frac{3}{32} & \frac{9}{32} & & & & \\
\frac{12}{13} & \frac{1932}{2197} & -\frac{7200}{2197} & \frac{7296}{2197} & & & \\
1 & \frac{439}{216} & -8 & \frac{3680}{513} & -\frac{845}{4104} & & \\
\frac{1}{2} & -\frac{8}{27} & 2 & -\frac{3544}{2565} & \frac{1859}{4104} & -\frac{11}{40} & \\
\hline
u_{l+1} & \frac{25}{216} & 0 & \frac{1408}{2565} & \frac{2197}{4104} & -\frac{1}{5} & 0 \\
\hline
\hat{u}_{l+1} & \frac{16}{135} & 0 & \frac{6656}{12825} & \frac{28561}{56430} & -\frac{9}{50} & \frac{2}{55}
\end{array}$$

is choosen as the basis for numerical solving ODE in most of software packages.

For solving FDE it should be supplemented by the interpolation and the extrapolation of the fifth degree.

Practically the same accuracy has the embedded seven-stage *Dorman-Prince method* 5(4):

$$\begin{array}{c|ccccccc}
0 & & & & & & & \\
\frac{1}{5} & \frac{1}{5} & & & & & & \\
\frac{3}{10} & \frac{3}{40} & \frac{9}{40} & & & & & \\
\frac{4}{5} & \frac{44}{45} & -\frac{56}{15} & \frac{32}{9} & & & & \\
\frac{8}{9} & \frac{19372}{6561} & -\frac{25360}{2187} & \frac{64448}{6561} & -\frac{212}{729} & & & \\
1 & \frac{9017}{3168} & -\frac{355}{33} & \frac{46732}{5247} & \frac{49}{176} & -\frac{5103}{18656} & & \\
1 & \frac{35}{384} & 0 & -\frac{500}{1113} & \frac{125}{192} & -\frac{2187}{6784} & \frac{11}{84} & \\
\hline
u_{l+1} & \frac{35}{384} & 0 & \frac{500}{1113} & \frac{125}{192} & -\frac{2187}{6784} & \frac{11}{84} & 0 \\
\hline
\hat{u}_{l+1} & \frac{5179}{57600} & 0 & \frac{7571}{16695} & \frac{393}{640} & -\frac{92097}{339200} & \frac{187}{2100} & \frac{1}{40}
\end{array}$$

which has the main method of the fifth order, and the error estimation is realized by the method of the fourth order. The method is of the class of continuous methods, because it allows one to calculate

$$u(t_l + r\Delta) = u_l + \Delta \sum_{i=1}^{6} \sigma_i(r) h_i, \quad 0 \leq r \leq 1,$$

With the coefficients defined by formulas

$$\sigma_1(r) = r(1+r(-1337/480+r(1039/360+r(-1163/1152)))),$$

$$\sigma_2(r) = 0,$$

$$\sigma_3(r) = 100r^2 \frac{1054/9275 + r(-4682/27825 + r(379/5565))}{3},$$

$$\sigma_4(r) = -5r^2 \frac{27/40 + r(-9/5 + r(83/96))}{2},$$

$$\sigma_5(r) = 18225r^2 \frac{-3/250 + r(22/375 + r(-37/600))}{848},$$

$$\sigma_6(r) = -22r^2 \frac{-3/10 + r(29/30 + r(-17/24))}{7}.$$

In such a way the interpolation of the fifth order and extrapolation are realized, respectively, that allows one numerically solve not only ODE, but also and FDE.

23.4 Influence of the approximate calculating functionals of the righ-hand side of FDEs

Typical example of FDE with distributed delays is a system which right-hand side contains integrals of the prehistory or functions of prehistory. In this case necessary approximately calculate integrals that can influence on the accuracy or a method. The most convenient method of calculating the discussed discrete scheme are the composite quadrature formulas.

The parameter which characterizes the accuracy is the value of the discretization step (in case of variable step – the maximal value of the step).

In this section taking as the example the explicit Runge-Kutta-like methods with variable step we derive estimations of the global error taking into account approximate calculation of FDE righ-hand side.

Assume that one can calculate for the system (23.1) not the exact values of the functional $f(t, x(t), x_t(\cdot))$, but only its approximations $\hat{f}(t, x(t), x_t(\cdot))$, however we also assume that the functional $\hat{f}$ as well as the functional f, satisfies the conditions of the section 17, namely: it satisfies the Lipschits conditions with respect to the second–third variables and is continuous on shift. Similar to the section 23.1 set on $[t_0, t_0 + \theta]$ non-uniform grid $t_{l+1} = t_l + \Delta_l$, $l = 0, 1, \ldots, N-1$, where $\Delta_l > 0$ is the step. Denote $\Delta_{max} = \max_l \Delta_l$, $\Delta_{min} = \min_l \Delta_l$.

Obviously there exists $K > 0$ such that $\Delta_{max} \leq K\Delta_{min}$, consider the discrete model of system (23.1), denoting the approximation of the solution $x(t_l) = x_l$ at point t_l by $u_l \in \mathbb{R}^n$.

Assume that on the set of the extended discrete model prehistories $\{u_i\}_l^+$ an interpolation-extrapolation operator IE, with properties described in the section 23.1, is defined.

Definition 23.5 *The explicit k-stage Runge-Kutta-like (ERK) method with variable step size and the approximate calculation of the functional f is the discrete model:*

$$u_0 = x_0, \quad u_{l+1} = u_l + \Delta_l \hat{\Psi}(t_l, u(\cdot)), \quad l = 1, \ldots, N-1, \tag{23.25}$$

$$u(\cdot) = IE(\{u_i\}_l^+), \quad \hat{\Psi}(t_l, u(\cdot)) = \sum_{i=1}^{k} \sigma_i \hat{h}_i, \tag{23.26}$$

$$\hat{h}_1 = \hat{h}_1(t_l, u(\cdot)) = \hat{f}(t_l, u_l, u_{t_l}(\cdot)), \tag{23.27}$$

$$\hat{h}_i = \hat{h}_i(t_l, u(\cdot)) = \hat{f}(t_l + a_i\Delta_l, u_l + \Delta_l \sum_{j=1}^{i-1} b_{ij}\hat{h}_j, u_{t_l + a_i\Delta_l}(\cdot)), \tag{23.28}$$

$$|a_i| \leq a, \ i = 1, \ldots, k.$$

Let $x_l = x(t_l)$, $l = 0, 1, \ldots, N$ be the values of the exact solution $x(t)$ of the problem (23.1) – (23.3) at points t_l. The *approximation of the functional f has the approximation error p on the exact solution*, if there exists a constant C_f, such that for all $l = 0, 1, ..., N$ the following inequality is satisfied

$$\|f(t_l, x_l, x_{t_l}(\cdot)) - \hat{f}(t_l, x_l, x_{t_l}(\cdot))\| \leq C_f \Delta_{max}^p. \tag{23.29}$$

The *residual* of ERK-like method on $x(t)$ at points t_l are functionals

$$g_l(x(\cdot)) = \frac{x_{l+1} - x_l}{\Delta_l} - \Psi(t_l, x(\cdot)), \tag{23.30}$$

where

$$\Psi(t_l, x(\cdot)) = \sum_{i=1}^{i-1} \sigma_i h_i,$$

$$h_i = f(t_l + a_i\Delta_l, x_l + \Delta_l \sum_{j=1}^{i-1} b_{ij} h_j, x_{t_l + a_i\Delta_l}(\cdot)).$$

The *ERK-like method* has an approximation order with p-approximate calculating of the functional f on a class of problems F, if for every $x(\cdot) \in X(F)$ there exists a constant $C_{x(\cdot)}$ such that

$$\|g_l(x(\cdot))\| \leq C_{x(\cdot)} \Delta^p_{max}, \quad l = 0, \ldots, N-1. \qquad (23.31)$$

The *residual with approximate calculating the functional f of ERK-method on $x(t)$ at points t_l are functionals*

$$\hat{g}_l(x(\cdot)) = \frac{x_{l+1} - x_l}{\Delta_l} - \hat{\Psi}(t_l, x(\cdot)), \qquad (23.32)$$

where

$$\hat{\Psi}(t_l, x(\cdot)) = \sum_{i=1}^{i-1} \sigma_i \hat{h}_i,$$

$$\hat{h}_i = \hat{f}(t_l + a_i\Delta_l, x_l + \Delta_l \sum_{j=1}^{i-1} b_{ij} \hat{h}_j, x_{t_l + a_i\Delta_l}(\cdot)).$$

The ERK-method has an approximation order p with approximate calculating of the functional f on a class of problems F, if for every $x(\cdot) \in X(F)$ there exists a constant $\hat{C}_{x(\cdot)}$ such that

$$\|\hat{g}_l(x(\cdot))\| \leq \hat{C}_{x(\cdot)} \Delta^p_{max}, \quad l = 0, \ldots, N-1. \qquad (23.33)$$

Lemma 23.2 *Let the ERK-method has the approximation order p_1 and, the approximation of the functional f has the approximation order p_2, then the ERK-method has the approximation order with approximate calculating and the functional equals $p = \min\{p_1, p_2\}$.*

Proof. First, by the induction method we prove that for every $i = 1, \ldots, k$ there exist constants C_{h_i}, such that

$$\|\hat{h}_i(t_l, x(\cdot)) - h_i(t_l, x(\cdot))\| \leq C_{h_i} \Delta_{max}^{p_2}. \tag{23.34}$$

The basis of the induction.

$$\|\hat{h}_1(t_l, x(\cdot)) - h_1(t_l, x(\cdot))\| =$$

$$= \|\hat{f}(t_l, x_l, x_{t_l}(\cdot)) - f(t_l, x_l, x_{t_l}(\cdot))\| \leq C_f \Delta_{max}^{p_2}.$$

The induction step. Assume that the estimation (23.34) is valid for all indices $< i$. Prove it for i.

$$\|\hat{h}_i(t_l, x(\cdot)) - h_i(t_l, x(\cdot))\| =$$

$$= \|\hat{f}(t_l + a_i\Delta_l, x_l + \Delta_l \sum_{j=1}^{i=1} b_{ij}\hat{h}_j, x_{t_l+a_i\Delta_l}(\cdot)) -$$

$$- f(t_l + a_i\Delta_l, x_l + \Delta_l \sum_{j=1}^{i-1} b_{ij}h_j, x_{t_l+a_i\Delta_l}(\cdot))\| \leq$$

$$\leq \|\hat{f}(t_l + a_i\Delta_l, x_l + \Delta_l \sum_{j=1}^{i-1} b_{ij}\hat{h}_j, x_{t_l+a_i\Delta_l}(\cdot)) -$$

$$- f(t_l + a_i\Delta_l, x_l + \Delta_l \sum_{j=1}^{i-1} b_{ij}\hat{h}_j, x_{t_l+a_i\Delta_l}(\cdot))\| +$$

$$+ \|f(t_l + a_i\Delta_l, x_l + \Delta_l \sum_{j=1}^{i-1} b_{ij}\hat{h}_j, x_{t_l+a_i\Delta_l}(\cdot)) -$$

$$- f(t_l + a_i\Delta_l, x_l + \Delta_l \sum_{j=1}^{i-1} b_{ij}h_j, x_{t_l+a_i\Delta_l}(\cdot))\| \leq$$

$$\leq C_f \Delta_{max}^{p_2} + L\Delta_l b \sum_{j=1}^{i-1} C_{h_j} \Delta_{max}^{p_2}.$$

Then we have the estimation (23.34), from which obtain

$$\|\hat{g}_l(x(\cdot))\| \leq \|g_l(x(\cdot))\| + \|\hat{\Psi}(t_l, x(\cdot)) - \Psi(t_l, x(\cdot))\| \leq$$

$$\leq C_{x(\cdot)}\Delta_{max}^{p_1} + k\sigma \sum_{i=1}^{k} C_{h_i}\Delta_{max}^{p_2},$$

which proves the proposition. □

The ERK-method (23.25) – (23.28) for the problem (23.1) – (23.3) *converges*, if $x_l - u_l \to 0$ as $\Delta_{max} \to 0$ for all $l = 0, 1, \ldots, N$, and the method *converges with the order* p, if there exists a constant C_u, such that

$$\|x_l - u_l\| \leq C_u \Delta_{max}^p, \quad l = 0, 1, \ldots, N. \tag{23.35}$$

Theorem 23.4 *If ERK-like method (23.25) – (23.28) has the approximation order $p_1 > 0$, the approximation of the functional f has approximation order p_2 and the interpolation–extrapolation operator IE has the order $p_3 >$ 0. Then ERK-like method converges and has the convergence order $p = \min\{p_1, p_2, p_3\}$.*

Proof. Denote $\delta_l = u_l - x_l$, then from (23.25) follows $\delta_0 = 0$.

For $l = 0, \ldots, N-1$ we have

$$\delta_{l+1} = \delta_l + \Delta_l[\hat{\Psi}(t_l, IE(\{u_i\}_l^+)) - \hat{\Psi}(t_l, IE(\{x_i\}_l^+))] +$$

$$+[x_l - x_{l+1} + \Delta_l\hat{\Psi}(t_l, x(\cdot))] + \Delta_l[\hat{\Psi}(t_l, I(\{x_i\}_l^+)) - \hat{\Psi}(t_l, x(\cdot))].$$

Applying to the function $\hat{\Phi}(\{u_i\}_l^+) = \hat{\Psi}(t_l, IE(\{u_i\}_l^+))$ the Lipschits condition with the constant $L_{\hat{\Phi}}$ with respect to the extended model prehistory similar to the corollary 23.1, taking into account (23.33), and also (23.28), we have

$$\|\delta_{l+1}\| \leq \|\delta_l\| + \Delta_l L_{\hat{\Phi}} \max_{t_l - 2\tau \leq t_i \leq t_l} \|\delta_i\| + \hat{C}_{x(\cdot)}\Delta_l^{p+1} + L_{\hat{\Psi}} C_{IE}\Delta_l^{p+1}. \tag{23.36}$$

The remaining part of the proof repeats the proof of the theorem 23.1. □

This proposition gives the accuracy of calculating functionals of the right-hand sides of FDEs, in order to obtain the appropriate convergence order of the method. For example in the software of the *Time-delay system Toolbox.* the Runge-Kutta-like methods 4(5) orders with the interpolation and extrapolation by the degenerate splines of the fourth order are realized.

To save in software convergence the order of the method for calculating the integrals of systems with distributed delays realized that the compound Simpson method [28] is used.

23.5 Test problems

In the section we consider several simple examples which can be used for testing numerical methods and software. In some examples it is possible to exact analytical solutions. This allows one the possibility of comparing the approximate and exact solutions and by value of the error (as well as computational time and necessary memory) to form an opinion about the effectiveness of different numerical algorithms, methods of interpolation and extrapolation discrete model prehistory, methods of forming the steps.

In the last examples of this section exact solutions are not known, however we know some qualitative characteristics of the solution behaviour: asymptotics, limit cycles. Realized numerical computations allow one to supplement knowledge about qualitative properties of solutions of these and other models.

Remember that the numerical computations were realized by Runge-Kutta-like methods 4(5) with automatic step size control, with interpolation by degenerate splines of the fourth order and extrapolation by the continuation.

This method in the best manner has proved itself in simulations of most FDE problems. However there may be *stiff* problems, so it was chosen as the basis for the software

[28] Which has the fourth accuracy order

of the Time-Delay System Toolbox presented in the section 24.

Test 1.

Consider the one-dimensional FDE

$$\dot{x}(t) = x(t - \tau),$$

where $(t \geq 0, \quad \tau = 1)$ with the initial condition

$$x(s) = 1, -1 \leq s \leq 0.$$

The exact solution of this problem we constructed by the step method in the section 13.2 (see example 13.1).

On figure 1 one can find graphs of the exact solution (*solid line*) and the approximate solution (*rings*, corresponding to the grid). These notation for the exact and approximate solutions are used on all other figures of the corresponding tests.

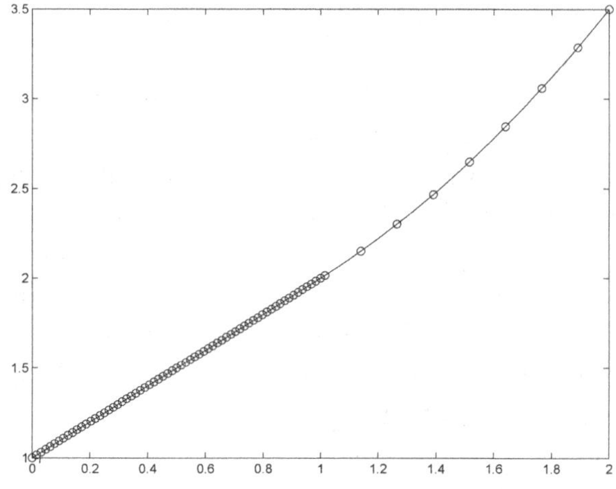

Fig. 1

Test 2.

Consider an equation with variable delay

$$\dot{x}(t) = x(t - (e^{-t} + 1)) + \cos t - \sin(t - e^{-t} - 1),$$

(see example 13.2).

On the figure 2 one can see graphs of the exact and approximate solutions.

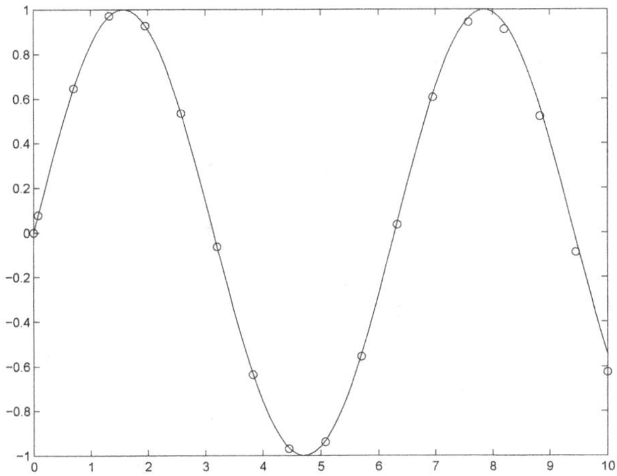

Fig. 2

Test 3.

Consider an equation with distributed delay

$$\dot{x}(t) = \int_{-1}^{0} x_t(s)ds,$$

(see example 13.3).

On the figure 3 one can see graphs of the exact and approximate solutions.

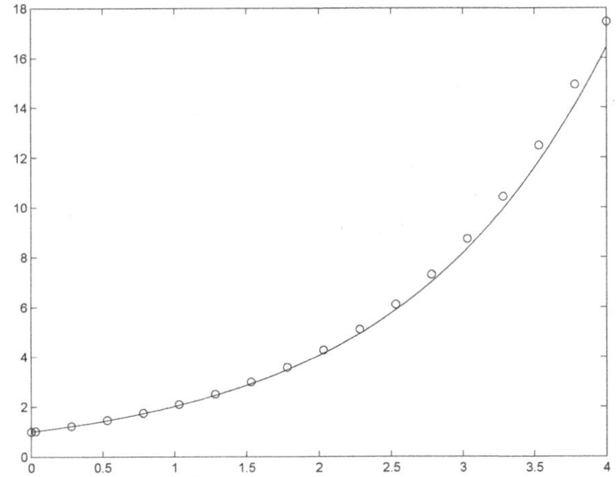

Fig. 3

Test 4.

Linear second order system with discrete variable and distributed delays

$$\begin{cases} \dot{x}_1(t) = -\sin(t)x_1(t) + x_1(t-\frac{t}{2}) - \\ \qquad - \int\limits_{-\frac{t}{2}}^{0} \sin(t+s)x_1(t+s)ds - e^{\cos(t)}, \\ \dot{x}_2(t) = \cos(t)x_2(t) + x_2(t-\frac{t}{2}) + \\ \qquad + \int\limits_{-\frac{t}{2}}^{0} \cos(t+s)x_2(t+s)ds - e^{\sin(t)} \end{cases}$$

with the initial conditions

$$x_1(t) = e^{\cos(t)},\ x_2(t) = e^{\sin(t)} \text{ for } t \le 0.$$

The system is considered on the interval $[0, 2\pi]$ and has the exact solution, defined by the same formulas as the initial functions. On figure 4 one can see graphs of two coordinates of the exact and approximate solutions.

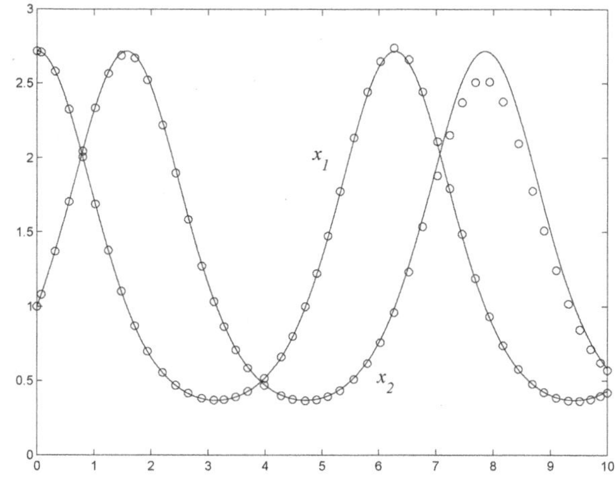

Fig. 4

Test 5. The nonlinear of the second order system with the distributed delay

$$\begin{cases} \dot{x}_1(t) = -\frac{1}{2}\int\limits_{-\pi}^{0} x_1(t+s)ds + \frac{2x_1(t)-\frac{\pi}{2}x_2(t)}{\sqrt{x_1^2(t)+x_2^2(t)}}, \\ \dot{x}_2(t) = -\frac{1}{2}\int\limits_{-\pi}^{0} x_2(t+s)ds + \frac{2x_2(t)+\frac{\pi}{2}x_1(t)}{\sqrt{x_1^2(t)+x_2^2(t)}} \end{cases}$$

The initial conditions:

$$x_1(t) = t\cos(t),\; x_2(t) = t\sin(t) \quad \text{for } t \leq 1.$$

On the interval $[1, 20]$ this system has the exact solution, which is defined the same formulas as the initial functions. In figure 5 one can see graphs of two coordinates of the exact and approximate solutions. In figure 6 — the corresponding trajectories on the phase plane are shown.

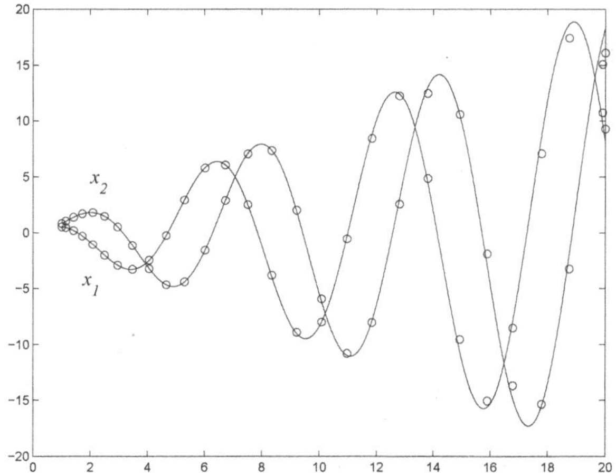

Fig. 5

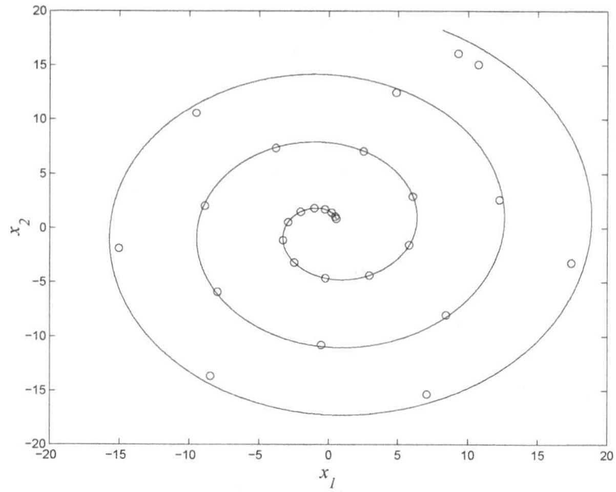

Fig. 6

Test 6.

The linear third order system with the constant delay

$$\begin{cases} \dot{x}_1(t) = \frac{2}{\pi}(x_1(t) + x_2(t - \frac{\pi}{2})) - x_1(t - \frac{\pi}{2}) - \frac{\pi}{2}\frac{x_2(t)}{x_3(t)}, \\ \dot{x}_2(t) = \frac{2}{\pi}(x_2(t) - x_1(t - \frac{\pi}{2})) - x_2(t - \frac{\pi}{2}) + \frac{\pi}{2}\frac{x_1(t)}{x_3(t)}, \\ \dot{x}_3(t) = \frac{\sqrt{x_1^2(t-\frac{\pi}{2}) + x_2^2(t-\frac{\pi}{2})}}{x_3(t-\frac{\pi}{2})} \end{cases}$$

The initial conditions

$$x_1(t) = t\cos(t),\ x_2(t) = t\sin(t),\ x_3 = t \ \text{ for } t \leq \pi.$$

On the interval $[\pi, 20]$ the system has the exact solution, defined by the same formulas as the initial functions. In figure 7 one can see graphs of three coordinates of the exact and approximate solutions. In figure 8 — the corresponding trajectories in the phase space are shown.

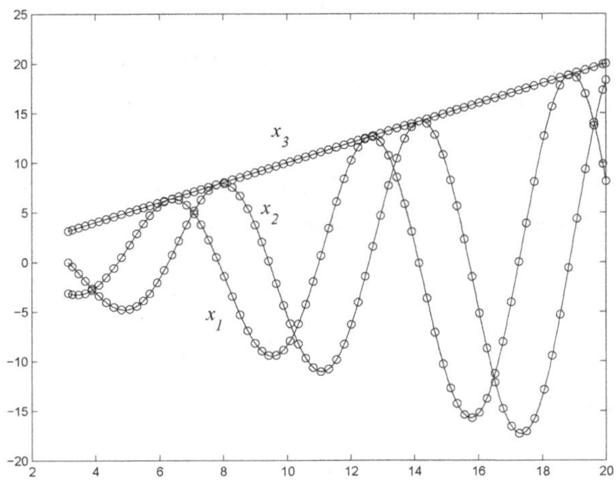

Fig. 7

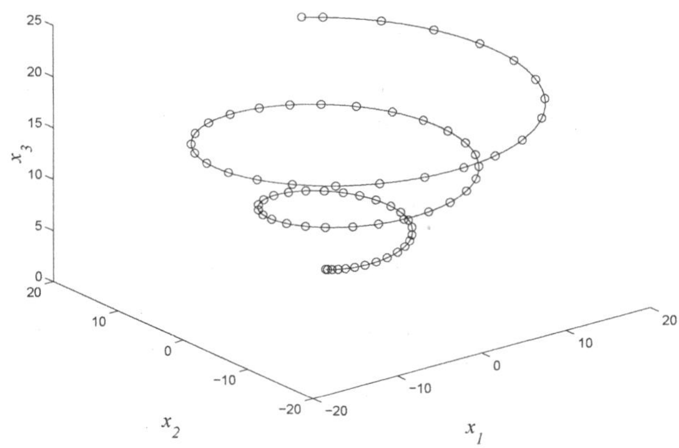

Fig. 8

Test 7.
The stiff system

$$\begin{cases} \dot{x}_1(t) = \lambda_1 x_1(t) + x_2(t) + x_1(t-1) - \frac{e^{\lambda_2(t-1)}}{\lambda_2-\lambda_1} - e^{\lambda_1(t-1)}, \\ \dot{x}_2(t) = \lambda_2 x_2(t) \end{cases}$$

with the initial conditions

$$x_1(t) = \frac{e^{\lambda_2 t}}{\lambda_2 - \lambda_1} + e^{\lambda_1 t}, \ x_2(t) = e^{\lambda_2 t} \ \text{ when } t \leq 0.$$

($\lambda_1 = -100,\ \lambda_2 = -1$) has the exact solution, defined by the same formulas as the initial functions.

Stiffness in the system is understood in the following (finite-dimensional sense).

Let $x(t)$ be the solution of problem (23.1) – (23.3). Calculate the Jacobi matrix of the functional f along the solution

$$A(t) = \frac{\partial f(t, x(t), x_t(\cdot))}{\partial x}$$

and consider the system of linear homogeneous differential equations

$$\dot{x}(t) = A(t)x(t),\ t \in [t_0, t_0 + \theta].$$

Denote by $\lambda_i(t),\ i = 1, ..., n$ the eigenvalues of the matrix $A(t)$.

Definition 23.6 *The problem (23.1) – (23.3) is the stiff problem, if*

1) Re $\lambda_i(t) < 0$ *for all* $t \in [t_0, t_0 + \theta],\ i = 1, ..., n$;

2) number

$$\sup_{t\in[t_0,t_0+\theta]} \{\max_{1\le i\le n} |Re\ \lambda_i(t)|/ \min_{1\le i\le n} |Re\ \lambda_i(t)|\}$$

is large.

In figure 9 one can see graphs of two coordinates of the exact and approximate solutions.

Testing a stiff problem by explicit methods[29] can lead to the unsatisfactory results: implicit FDE numerical methods can then be applied. For comparison on figure 10 we present results of computer simulating obtained by implicit method with constant step.

[29]even with automatic step size control

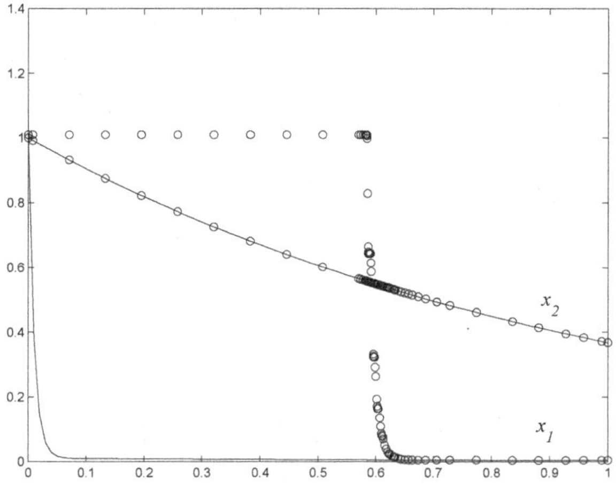

Fig. 9

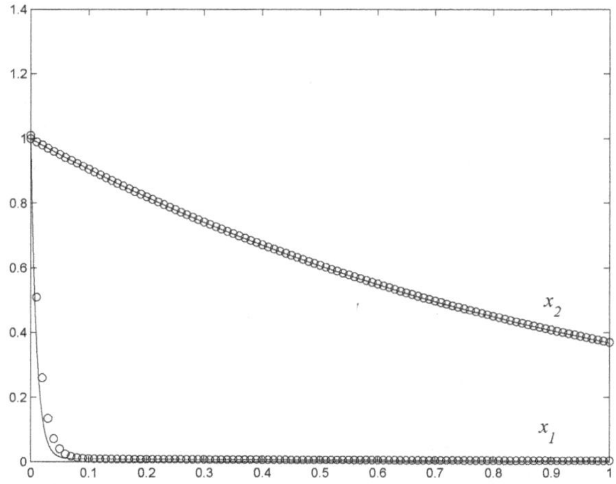

Fig. 10

Test 8. *Lotka–Volterra system* with distributed delay

$$\begin{cases} \dot{x}_1(t) = [\varepsilon - \gamma_1 x_2(t) - \int\limits_{-\tau}^{0} F_1(s) x_2(t+s) ds] x_1(t) \\ \dot{x}_2(t) = [-\varepsilon + \gamma_2 x_1(t) + \int\limits_{-\tau}^{0} F_2(s) x_1(t+s) ds] x_2(t). \end{cases}$$

On figure 11 one can see graphs of two coordinates of the exact and approximate solutions. On figure 12 there is the corresponding approximate trajectory in phase space.

the calculation were realized with theparameters: $\tau = 1$, $\varepsilon = 2$, $\gamma_1 = 1$, $\gamma_2 = 1$, $F_1(s) = \sin s$, $F_2(s) = \sin s$, initial conditions $x_1(s) \equiv 1$, $x_2(s) \equiv 1$ for $-\tau \leq s \leq 0$.

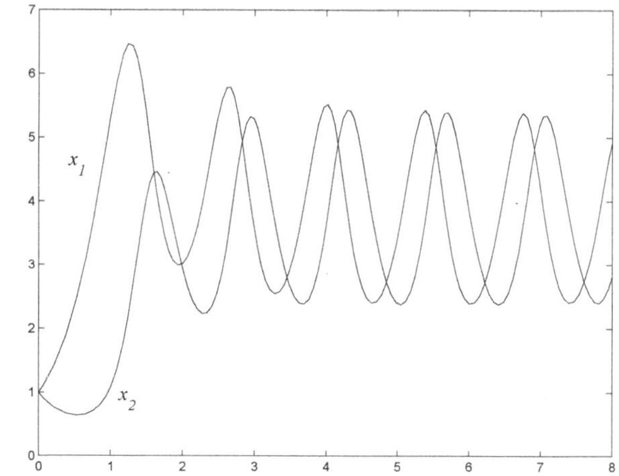

Fig. 11

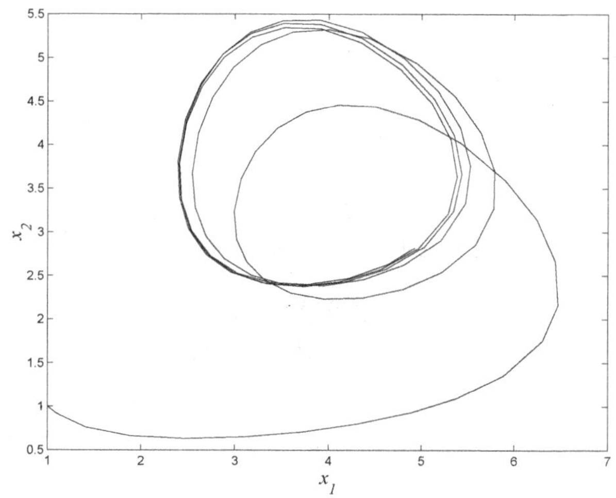

Fig. 12

Test 9. Equation with two delays

$$\dot{N}(t) = N(t)[\gamma - a_0 N(t) - a_1 N(t - \tau_1) + bN(t - \tau_2)],$$

where $\gamma > 0$, $a_0 \geq 0$, $a_1 > 0$, $b > 0$, nonnegative values of delays τ_1, τ_2, are, generally speaking, incommensurate. The example belongs to I. Gyori, who announced to us the following properties of the solution:

1) If $a_0 + a_1 > b$ and $a_1\tau_1$ is sufficiently small, then any positive solution $N(t)$ tends to a constant as $t \to \infty$;

2) if $a_1 > b$ and $\tau_1 > \tau_2$, then there exist solutions which $N(t) \to \infty$ as $t \to \infty$.

The following question is open: if $a_0+a_1 > b$ and $\tau_1 < \tau_2$, then are there solutions which $N(t) \to \infty$ as $t \to \infty$?

On figures 13 and 14 the graphs of the approximate solutions, corresponding these two cases are presented. We used the following parameters:

case 1) $\gamma = 1$, $a_0 = 1$, $a_1 = 2$, $b = 1$,$\tau_1 = 1$, $\tau_2 = 1$;

case 2) $\gamma = 0.8$, $a_0 = 0$, $a_1 = 2$, $b = 1$, $\tau_1 = \ln 10$, $\tau_2 = \ln 2$;

initial conditions $N(s) \equiv 1$ for $-\tau_1 \leq s \leq 0$.

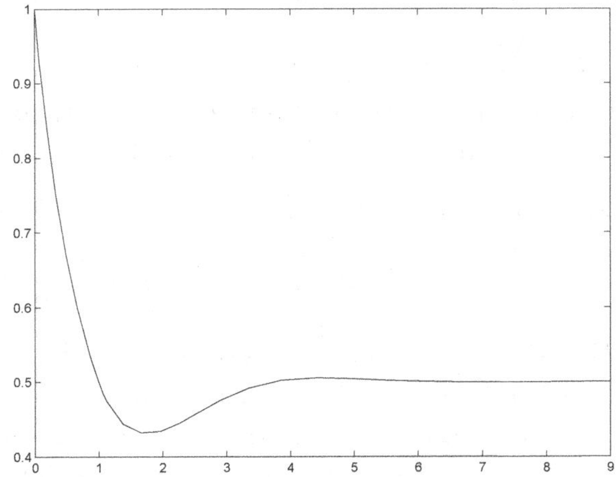

Fig. 13

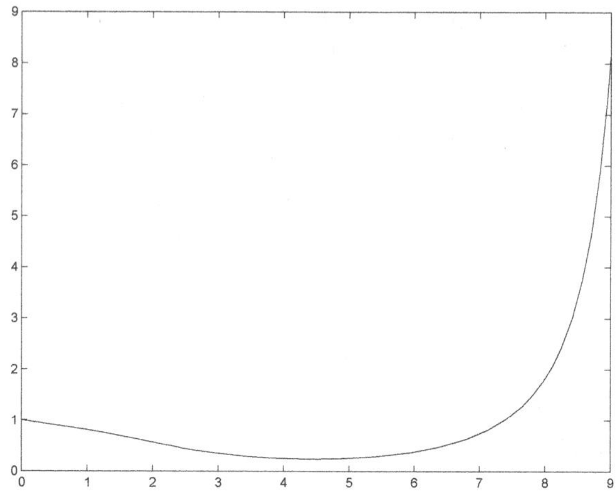

Fig. 14

Test 10.

Oscillation of the pendulum with vibrating suspension [33]

$$\ddot{x}(t)+2q\varepsilon\dot{x}(t)+k^2\varepsilon^2\sin x(t-h(t))-\varepsilon\cos t\cos x(t-\delta(t))-$$

$$-\varepsilon r\sin(t+\kappa)\sin x(t-\delta(t))=0,$$

where x is the angle of deviating of the pendulum from the vertical line. Point of suspension moves according to the law:

$u(t) = a\cos\omega t$, $v(t) = b\sin(\omega t+\kappa)$, a, b, ω, κ — constants, l — length of the pendulum, g — acceleration, λ_1— coefficient of damping, $k^2 = \frac{gl}{a^2\omega^2}$, $q = \frac{\lambda_1 k}{2\sqrt{g/l}}$.

It is assume that the following conditions are valid

$$a/l = \varepsilon << 1,\ b/l = \varepsilon r << 1,\ \lambda_1 = \varepsilon\lambda,\ \omega > \sqrt{gl}/a.$$

Delays $h(t)$ and $\delta(t)$ are continuous differentiable, bounded ($0 \le h(t) \le R$, $0 \le \delta(t) \le R$) and 2π–periodic functions.

In the equations movement is taken into account which delays the effective restore force $h(t)$ and delays the impulse of the energy source, where oscillation originates $\delta(t)$.

In figure 15 one can see graphs of two coordinates approximate solution, in figure 16 — the corresponding approximate trajectory in the phase space.

Simulation was realized under the following parameters: $l = 1$, $\lambda_1 = 1$, $a = 0.1$, $b = 0.01$, $\omega = 50$, $\kappa = 1$, $h(t) = 2\sin t + 5\pi/2$, $\delta(t) = 2\cos t + 5\pi/2$.

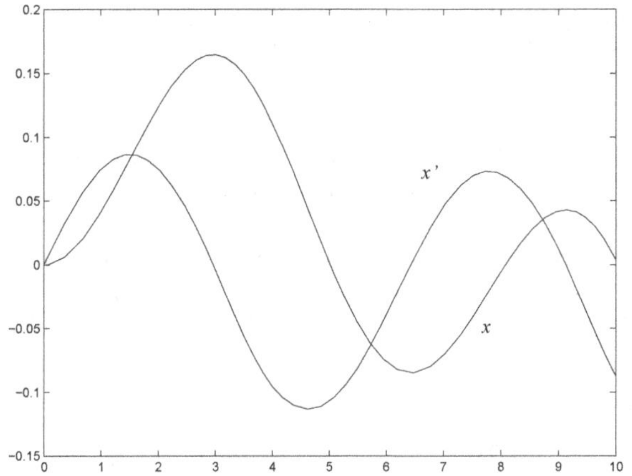

fig. 15

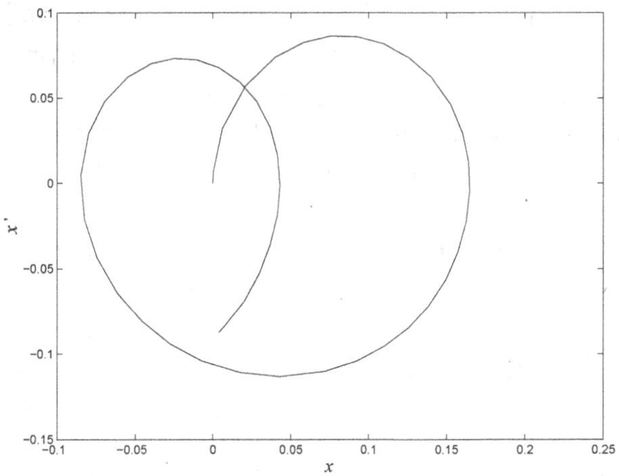

Fig. 16

24 Software package Time-delay System Toolbox

24.1 Introduction

Time-delay System Toolbox (for use with MATLAB) is the software package for computer simulation and analysis of systems with delays.

Programs of the *Time-delay System Toolbox* (*TDST*) provide support for:

- Numerical simulating linear and nonlinear systems with discrete and distributed delays;
- *Time-domain analysis* of linear systems with delays;
- Analytical design technique for linear quadratic stabilization problems with delays.

24.2 Algorithms

The numerical methods and control design algorithms, realized in *Time-Delay System Toolbox*, are direct generalization of the corresponding methods of ordinary differential equations (ODE) case theory, i.e. if delays disappear then the algorithms coincide with the corresponding numerical and control algorithms for ODEs. So one can also simulate and analyze finite-dimensional systems (ODE) using this toolbox.

Numerical algorithms. Simulation of systems with delays is realized on the basis of the numerical Runge-Kutta-like methods described in the previous section.

Time-domain analysis. Special functions of *TDST* allow one to realize *time-domain analysis* of linear systems with delays.

Stability. Stability is one of the most important characteristics of (linear) systems with delays. The special algorithm of stability verification of linear time-delay systems is realized in *TDST*.

Control algorithms. The specific methods of analytical constructing linear quadratic regulators for time-delay systems are realized in *TDST*. The methods are generalization by the time-delay case using the classic methods of analytical constructing regulators for finite-dimensional systems. Therefore, if delays disappear, the algorithms coincide with the standard algorithms for ODE.

24.3 The structure of the Time-delay System Toolbox

1. Numerical algorithms:

– Simulation of nonlinear systems with delays (**fde45**).
– Simulation of Linear systems with delays (**fde45lin**).
– Lotka-Volterra system with delays (**lv45**).

2. Time-domain analysis of linear systems with delays:

– Fundamental (transition) matrix of a linear system with delays (**fundmatr**).

– Impulse response of a linear system with delays (**impulsed**).

– Initial condition response of a linear system with delays (**initiald**).

– Arbitrary input response of a linear system with delays (**inputd**).

– Step input response of a linear system with delays (**stepd**).

3. Stability of Linear systems with delays:

– Stability of a linear system with delays in terms of the fundamental matrix (**checkstab**).

4. Control algorithms:

– A linear feedback control (**clsim**).
– Generalized LQR regulator (**clso**).
– Optimal gain (**lqdelay**).
– System of Generalized Riccati equatios (**gre**).

– Exponential matrix equation (**emed**).

5. *Delay operators:*
   – Discrete delay operator (**ydelay**).
   – Distributed delay operator (**int1**).
   – Distributed delay operator (**int2**).

6. *Miscellaneous algorithms.*

24.4 Descriptions of some programs

Program fde45

Purpose

Solution of nonlinear systems with delays.

Synopsis

[t,x]=fde45('xpfun',t0,tf,x0,initfun)
[t,x]=fde45('xpfun',t0,tf,x0,initfun,tol)

Description

Program **fde45** solves on the interval $[t_0, t_f]$ the system of nonlinear delay differential equations

$$\dot{x} = f(t, x, y(-\tau_1(t)), \ldots, y(-\tau_k(t)),$$

$$\int\limits_{-\tau_{k+1}(t)}^{0} \psi_1(t,s)\, y(s)\, ds, \ldots, \int\limits_{-\tau_{k+m}(t)}^{0} \psi_m(t,s)\, y(s)\, ds,$$

$$\int\limits_{-\tau_{k+m+1}(t)}^{0} \phi_1(y(s))\, ds, \ldots, \int\limits_{-\tau_{k+m+p}(t)}^{0} \phi_p(y(s))\, ds) \qquad (24.1)$$

with the initial conditions

$$\begin{cases} x(t_0) = x^0, \\ x(t_0 + s) = y^0(s), \quad -\tau \le s < 0, \end{cases} \tag{24.2}$$

where

$$\tau = \max\Big\{ \max_{t_0 \le t \le t_f} \tau_i(t)\,, \; i = \overline{1, k+m+p} \Big\}.$$

Function **fde45** returns the approximate solution of the problem (24.1)–(24.2) and automatically determines the time vector t. No plot is drawn on the screen. The matrix x contains the values of the approximate solution respectively evaluated at the time points t and has as many columns as the system dimension and one row for each element of t.

The discrete delay terms

$$y(-\tau_i)\,, \; i = \overline{1, k}\,,$$

and distributed delay terms

$$\int\limits_{-\tau_{k+1}(t)}^{0} \psi_i(t, s)\, y(s)\, ds\,, \; i = \overline{1, m},$$

$$\int\limits_{-\tau_{k+1}(t)}^{0} \phi_j(y(s))\, ds\,, \; j = \overline{1, p}\,,$$

are calculated by functions **ydelay**, **int1**, **int2** respectively.

Input parameters:

xpfun	—	String variable with the right-hand side f of system (24.1),
t0	—	Initial time t_0,
tf	—	Final time t_f,
x0	—	Initial vector x^0 (finite-dimensional component of the initial position),
initfun	—	String variable with the functional component of the initial position function $y^0(s)$, $\tau^* \le s < 0$,
tol	—	Tolerance (by default 1.e-4).

Limitation

$\tau_i(t)$, $i = \overline{1, k+m+p}$, must be non-negative.

Diagnostics

If **fde45** cannot perform the integration over the full time rang requested a **singularity likely**.

Example

Solve on the segment $[1, 20]$ the system with delays

$$\begin{cases} \dot{x}_1 = -\dfrac{1}{2}\displaystyle\int_{-\pi}^{0} y_1(s)\,ds + \dfrac{2\,x_1 - \frac{\pi}{2}\,x_2}{\sqrt{x_1^2 + x_2^2}}, \\ \dot{x}_2 = -\dfrac{1}{2}\displaystyle\int_{-\pi}^{0} y_2(s)\,ds + \dfrac{2\,x_2 + \frac{\pi}{2}\,x_1}{\sqrt{x_1^2 + x_2^2}} \end{cases} \tag{24.3}$$

with the initial vector

$$x^0 = \begin{pmatrix} \cos(1) \\ \sin(1) \end{pmatrix} \tag{24.4}$$

and the initial function

$$y^0(s) = \begin{pmatrix} s\,\sin(s) \\ s\,\cos(s) \end{pmatrix}, \quad 1 - \pi \le s < 1. \tag{24.5}$$

The first step towards simulating the system is to create an M-file containing the description of the right-hand side of the system. We can call it *system.m*:

```
function output=system(t,x,tt,xt,initfun);
dx1=int1(-pi,0,t,tt,xt,initfun,'1',1);
dx2=int1(-pi,0,t,tt,xt,initfun,'1',2);
x1=-1/2*dx1+(2*x(1)-pi/2*x(2))/(sqrt(x(1)^2+x(2)^2));
x2=-1/2*dx2+(2*x(2)+pi/2*x(1))/(sqrt(x(1)^2+x(2)^2));
output=[x1 x2];
```

Now to simulate the system (defined by *system.m*) over the interval $[1, 20]$ invoke **fde45**

```
[t,x]=fde45('system',1,20,[cos(1) sin(1)],'[s*cos(s);s*sin(s)]');
```

To draw the plot of the approximate solution one can use

```
subplot(211);
plot(t,x,'o');
hold on t=1:0.01:20;
plot(t,[t.*cos(t);t.*sin(t)]);
```

To draw the phase portrait of the system

```
subplot(212);
plot(x(:,1),x(:,2),'o');
hold on
plot(t*(cos(t)),t*(sin(t)));
```

On figure 17 circles denote an approximate solution and a solid line is used for the exact solution

$$x(t) = \begin{pmatrix} t \cdot \cos(t) \\ t \cdot \sin(t) \end{pmatrix}, \quad t \geq 1 - \pi\,,$$

of the system (24.3) – (24.5).

On the bottom part of figure 17 phase portraits of the exact and approximate solutions are presented.

Algorithm

fde45 realizes numerical Runge-Kutta-like algorithm with automatic step size control [30].

Program ydelay

Purpose

Calculation of the discrete delay terms of the delay differential equations.

Synopsis

[30] see previous section

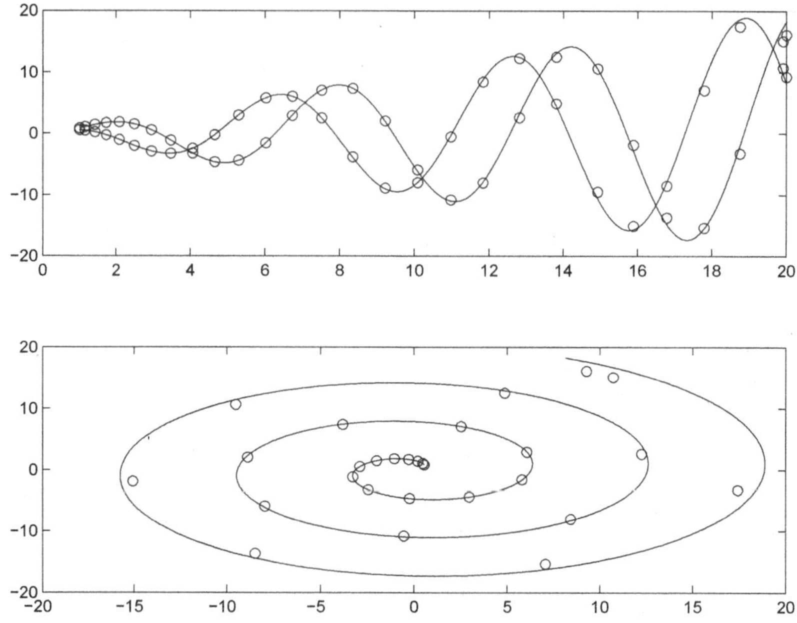

Figure 17

res = ydelay(arg,t,tt,xt,initfun)
res = ydelay(arg,t,tt,xt,initfun,num)

Description

Function **ydelay** calculates the value of the discrete delays component

$$y(-\tau(t)) ,$$

in the right-hand side of FDE. y is the state-delay vector of the system, $\tau(t)$ is the function delay given by user.

Function **ydelay** must be used in special user's m-file where the user describes the right-hand side of the system in order to solve it by **fde45**.

Input parameters:

arg — A function $-\tau(t)$, the argument of y,

num — An optional parameter: the number of coordinate of the vector x, returned by the function **ydelay** (in case of multidimension system).

Parameters **t, tt, yt, initfun** are not defined by user.

The output column-vector **res** is the result of calculation, it has as many elements as the system dimension, if parameter **num** is not given, and has just one element if parameter **num** is given.

Examples

1) Consider the system

$$\begin{cases} \dot{x}_1 = x_1 - 0.1\, y_2(-|\cos(t)|)\,, \\ \dot{x}_2 = x_2 + 0.1\, y_1(-|\sin(t)|)\,. \end{cases}$$

The first step towards simulating the system is to create M-file containing the description of the right-hand side of the system. We can call it *sys1.m*:

```
function output = sys1(t, x, tt, xt, initfun);
x1=x(1)-0.1*ydelay(-abs(cos(t)),t,tt,xt,initfun,2);
x2=x(2)+0.1*ydelay(-abs(sin(t)),t,tt,xt,initfun,1);
output=[x1 x2];
```

Now to simulate the system, defined by *sys1* over the segment $\pi \le t \le 6$ with the initial vector

$$x^0 = \begin{pmatrix} 0 \\ -1 \end{pmatrix}$$

and the initial function

$$y^0(s) = \begin{pmatrix} \sin(s) \\ \cos(s) \end{pmatrix}, \quad \pi - 1 \le s \le -\pi\,,$$

invoke **fde45**:

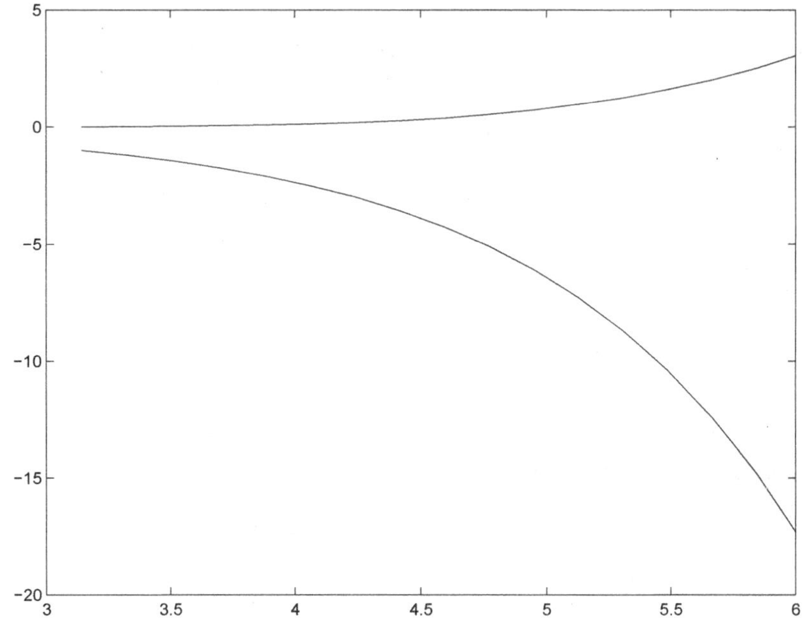

Fig. 18

```
[t,x]=fde45('sys1', pi, 6, [0; -1], '[sin(s); cos(s)]')
```

The result can be displayed by **plot(t,x)** (Fig. 18).

2) To find an approximate solution of the system

$$\begin{cases} \dot{x}_1 = -x_1 + x_1\, y_1(-1) + x_2\, y_2(-1), \\ \dot{x}_2 = -x_2 + x_1\, y_1(-1) + x_2\, y_2(-1) \end{cases}$$

we create the function *sys2.m* which describes the right-hand side of the system

```
function output = sys2(t, x, tt, xt, initfun);
output=-x + ydelay(-1,t,tt,xt,initfun)' * x;
```

To simulate the system over the segment $[0, 4]$ with the initial state vector

$$x^0 = \begin{pmatrix} 0 \\ 1 \end{pmatrix}$$

and the initial function

$$y^0(s) = \begin{pmatrix} \sin(s) \\ \cos(s) \end{pmatrix} \quad , \; -1 \leq s < 0 \, ,$$

invoke **fde45**:

```
[t,x]=fde45('sys2',0,4,[0; 1],'[sin(s); cos(s)]')
```

The graph can be plotted by **plot(t,x)** (Fig. 19).

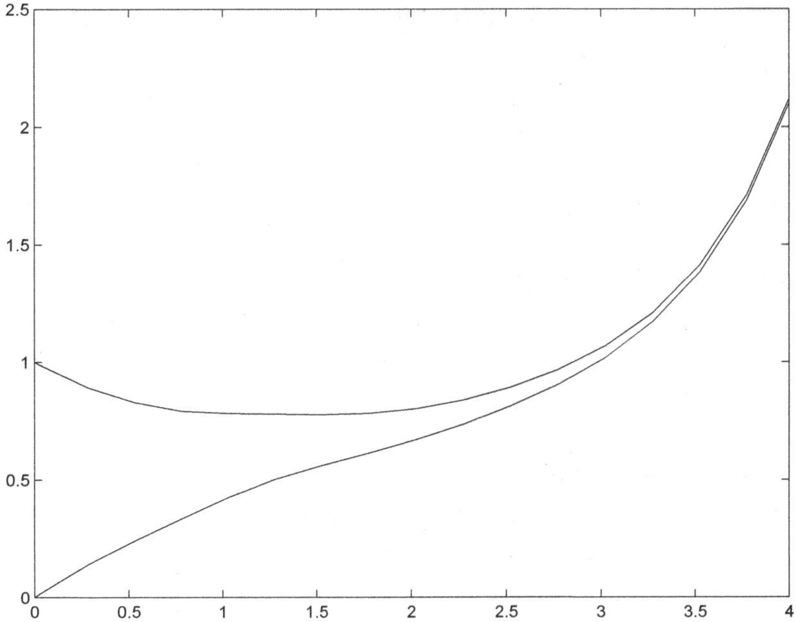

Fig. 20

Algorithm

ydelay utilizes the interpolation and extrapolation by splines of the third order.

Program int1

Purpose

Numerical integration.

Synopsis

res = int1(sb,se,t,tt,xt,initfun,F)
res = int1(sb,se,t,tt,xt,initfun,F,num)

Description

Function **int1** calculates the integral

$$\int_{-\tau(t)}^{0} F(t,s)\, y(s)\, ds \ ,$$

which describes the distributed delay in the right-hand side of FDE. Here y is the state-delay vector of the system, F and τ are the functions given by the user.

Function **int1** is used in the special user's m-file which describes the right-hand side of FDE in order to be solved by **fde45**.

Input parameters:

sb	—	Lower limit of integration: a function $-\tau(t)$,
se	—	Upper limit of integration (usually 0),
F	—	The string variable with the matrix F (depends on variables s and t),
num	—	Optional parameter: a number of coordinates of the vector x, returned by the function **int1** (in case of multidimensional system).

Parameters **t, tt, xt, initfun** are not defined by user.

Vector-column **res** contains results of numerical integration. Its dimension is as the dimension of the system, if the parameter **num** is not given, and it has just one element if the parameter **num** is given.

Examples

1) Consider a nonlinear system

$$\begin{cases} \dot{x}_1 = x_1 + \int\limits_{-1}^{0} (t+s)\, y_2(s)\, ds, \\ \dot{x}_2 = x_1\, y_1(-t/2). \end{cases}$$

The first step towards simulating the system is to create an m-file containing the right-hand side of the system. We can call it *sys.m*:

```
function output = sys(t, x, tt, xt, initfun);
x1=x(1)+int1(-1, 0, t, tt, xt, initfun, 's+t', 2);
x2=x(1)*ydelay(t/2, tt, xt,initfun, 1);
output =[x1 x2];
```

Now to simulate the system, defined by *sys.m* over the segment $0 \le t \le 5$ with the initial vector

$$x^0(0) = \begin{pmatrix} 0 \\ 0 \end{pmatrix}$$

and the initial function

$$y^0(s) = \begin{pmatrix} 0.2s \\ 0.2s \end{pmatrix}, \quad -1 \le s < 0,$$

invoke **fde45**:

```
[t,x]= fde45('sys', 0, 5, [0; 0], '[0.2*s; 0.2*s]')
```

The result can be displayed by **plot(t,x)** (Fig. 20).

2) To find a numerical solution of the equation

$$\dot{x} = y(-1) + \int\limits_{-t/2}^{0} \sin(s)\, y(s)\, ds$$

create m-file *fun.m*, containing the right-hand side of the equation:

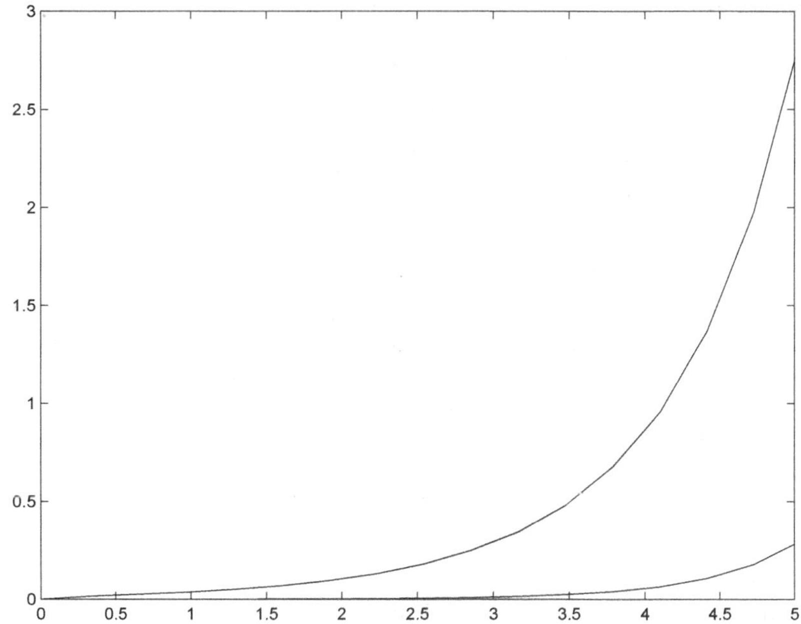

Fig. 20

```
function output = fun(t, x, tt, xt, initfun);
output=ydelay(t-1,tt,xt,initfun) + ...
     ... int1(-t/2,0,t,tt,xt,initfun,'sin(s)');
```

To simulate the equation over the segment $[0, 5]$ with the initial condition

$$\begin{cases} x^0(0) = 0\,, \\ y^0(s) = \sin(s)\,, \ -1 \le s < 0 \end{cases}$$

invoke **fde45**:

```
[t,x]=fde45('fun', 0, 5, 0, 'sin(s)')
```

The graph of the obtained solution can be plotted by command **plot(t,x)**.

Algorithm

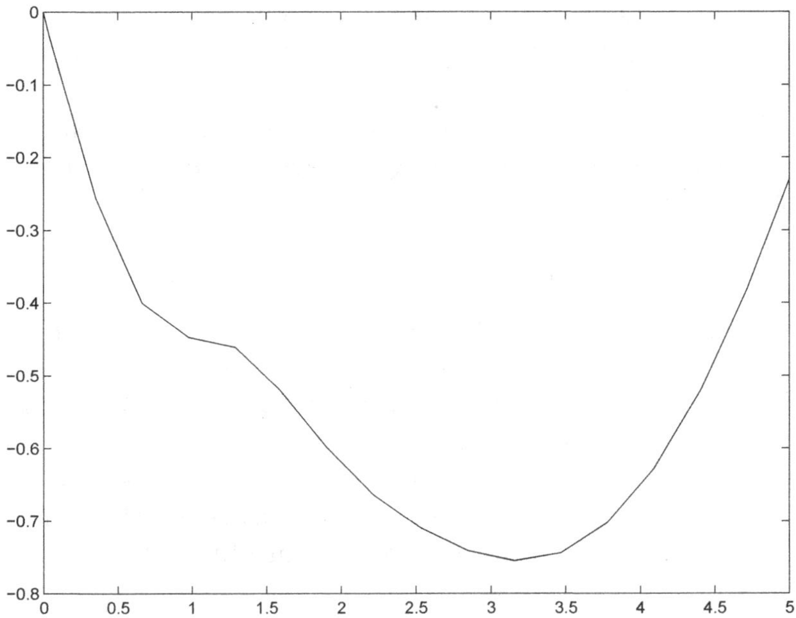

Fig. 21

In **int1** the Simpson method of integration is realized.

Program int2

Purpose

Numerical integration.

Synopsis

res = int2(sb,se,t,tt,xt,initfun,F)
res = int2(sb,se,t,tt,xt,initfun,F,num)

Description

Function **int2** calculates the integral

$$\int_{-\tau(t)}^{0} F(y(s))\, ds \;,$$

which describes the distributed delay in the right-hand side of FDE. Here y is the state-delay vector of the system, F and τ are the functions given by a user.

Function **int2** is used in special user's m-file which describes the right-hand side of the system in order to be solved by **fde45**.

Input parameters:

sb	—	Lower limit of integration: a function $-\tau(t)$,
se	—	Upper limit of integration (usually 0),
F	—	String variable with matrix F (depends on s),
num	—	Optional parameter: a number of coordinates of the vector x, returned by the function **int1** (in case of multidimensional system).

Parameters **t, tt, xt, initfun** are not defined by the user.

Vector-column **res** contains results of numerical integration. Its dimension is as the dimension of the system, if the parameter **num** is not given, and it has just one element if the parameter **num** is given.

Examples

1) Consider the nonlinear system

$$\begin{cases} \dot{x}_1 = \sin\left(x_2 + \int\limits_{-1}^{0} y_2^2(s)\,ds \right), \\ \dot{x}_2 = \cos\left(\int\limits_{-1}^{0} y_1^2(s)\,ds \right). \end{cases}$$

Create an M-file which describes the right-hand side of the system. We call it *sys.m*:

```
function output=sys(t, x, tt, xt, initfun);
x1=sin(x(2)+int2(-1, 0, t,tt,xt,initfun, 's^2', 2));
x2=cos(int2(-1, 0, t,tt,xt,initfun, 's^2', 1));
output=[x1 x2];
```

The first step towards simulating the system on segment $0 \le t \le 8$ is to create m-file containing the right side of the system. We can call it *sys.m*:

```
function output=sys(t,x,tt,xt,initfun);
x1=sin(x(2)+int2(-1,0,t,tt,xt,initfun,'s2',2))
x2=cos(int2(-1,0,tt,xt,initfun,'s2',1)
output=[x1,x2];
```

Now to simulate the system, defined by *sys.m*, over the segment $[0, 8]$ with the initial vector

$$x^0(0) = \begin{pmatrix} 0.3 \\ 0 \end{pmatrix}$$

and the initial function

$$y^0(s) = \begin{pmatrix} 0.3 + s \\ s \end{pmatrix}, \quad -1 \le s < 0,$$

invokes **fde45**:

```
[t,x]= fde45('sys', 0, 8, [0.3; 0], '[0.3+s; s]')
```

The result can be displayed by **plot(t,x)** (fig. 21).

2) To find an approximate solution of the equation

$$\dot{x} = x - \int_{-1}^{0} \sin(y(s))\, ds$$

it is necessary to create the function which describes the right side of the equation. We will call it *fun.m*:

```
function output = fun(t, x, tt, xt, initfun);
output=x-int2(-1, 0, t, tt, xt, initfun, 'sin(s)');
```

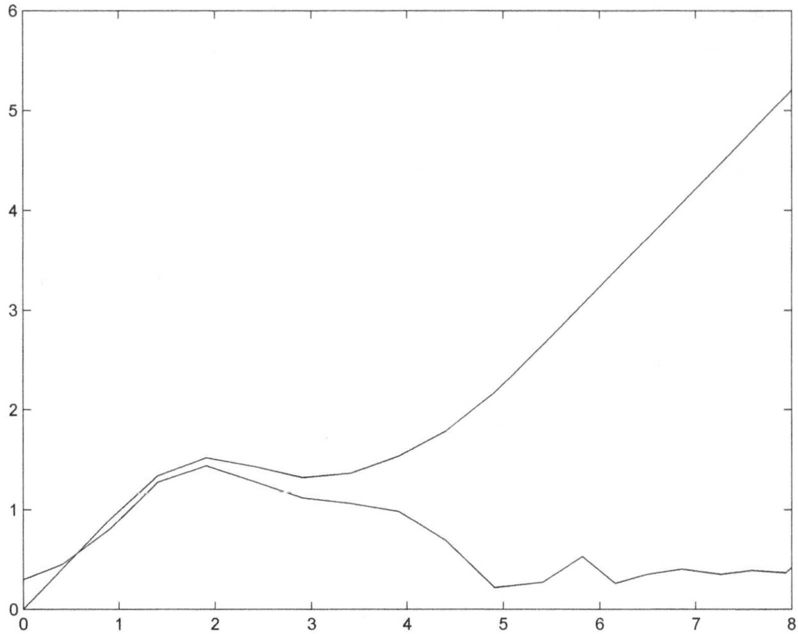

Fig. 21

To simulate the equation over the segment $[0, 8]$ with the initial condition

$$\begin{cases} x^0(0) = 0\,, \\ y^0(s) = \sin(s)\,, \ -1 \leq s < 0\,, \end{cases}$$

invoke **fde45** :

```
[t,x]=fde45('fun', 0, 8, 0, 'sin(s)')
```

The graph can be plotted by the command **plot(t,x)**(fig. 22).

Algorithm

In **int2** the compound Simpson method of integration is realized.

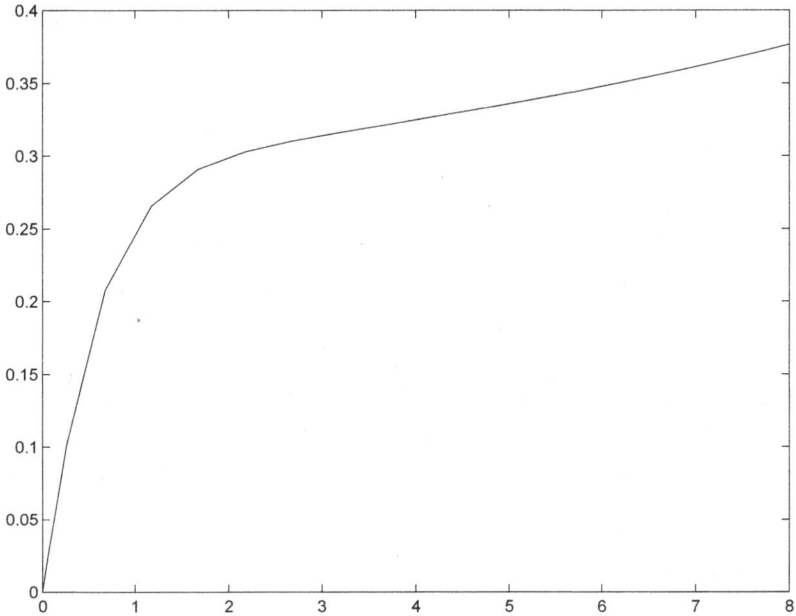

Fig. 22

Program spline3

Purpose

Numerical interpolation and extrapolation.

Synopsis

yi = spline3(x,y,xi)
yi = spline3(x,y,xi)

Description

The program **spline3** calculates values of a function at points of the array **xi**. Functions are given by the array **y**, corresponding to the array **x**.

spline3 is the auxiliary function for functions **int1**, **int2**, **ydelay** and is not defined by a use.

Input parameters:

x — The vector of coordinates at which functions are defining

y — Two-dimensional array of the function values, corresponding to x: every column of the array corresponds to a function,

xi — The vector of coordinates at which the functions should be calculated.

The two-dimensional array **yi** contains the value of functions, calculated at points **xi**. Every column of the array corresponds to a function. The number of columns in the array **yi** is equal to the number of functions. And the number of rows is equal to the number of the elements of the array **xi**.

Algorithm

In **spline3** the interpolation and extrapolation by the degenerate splines is realized.

Listing

```
  function yi=spline3(x,y,xi);
  %SPLINE3 Interpolation and extrapolation by degenerate splines.
  %    Given data vectors x and y and a new abscissa vector xi,
  %    the function yi=spline3(x,y,xi) uses interpolation and,
  %    if needed, extrapolation by the degenerate splines
  %    to find a vector yi corresponding to xi.

  % Interpolation and extrapolation for the simplest cases
  n=length(x);
if n==0
   return,
  end;
if n==1
     yi=ones(length(xi),1)*y;
     return
  end;
if n==2
     L1=(xi(:)-x(2))/(x(1)-x(2));
     L2=(xi(:)-x(1))/(x(2)-x(1));
     yi=[L1, L2]*y;
     return
  end;
if n==3
```

```
        L1=(xi(:)-x(2)).*(xi(:)-x(3))/((x(1)-x(2))*(x(1)-x(3)));
        L2=(xi(:)-x(1)).*(xi(:)-x(3))/((x(2)-x(1))*(x(2)-x(3)));
        L3=(xi(:)-x(1)).*(xi(:)-x(2))/((x(3)-x(1))*(x(3)-x(2)));
        yi=[L1 L2 L3]*y;
    end;
```

Part II

INVARIANT AND GENERALIZED DERIVATIVES OF FUNCTIONS AND FUNCTIONALS

In the second part we develop the theory of an invariant derivative of functions and investigate its relations with other function derivatives. In particular we show equivalence of the invariant derivative and the Sobolev generalized derivative.

In summary by means of the invariant derivatives we establish a mathematical relation between the Sobolev generalized derivative of functions and the generalized derivative of the distribution theory.

To understand the theory of this part the reader has to study the functional direction of i-smooth analysis.

25 The invariant derivative of functions

In this chapter we introduce the notion of the invariant derivative of functions and investigate its relation with other derivatives, in particular we prove its equivalence to the Sobolev generalized derivative.

25.1 The invariant derivative of functions

Further $[a, b] \subset R$ and we use the following notation:

$B[a, b]$ is a linear subspace of the space of integrable on $[a, b]$ functions,

$BLIN[a, b]$ is the space of linear on $B[a, b]$ functionals[31].

$< F, u >$ is the value of a linear functional $F \in BLIN[a, b]$ at an element $u \in B[a, b]$.

The invariant derivative of a function is defined by means of the invariant derivative of the linear functional generating this function.

The classic form of the presentation of a linear functional F by means of a function f has the integral form

$$< F, u > = \int_a^b f(x)u(x)dx, \tag{25.1}$$

[31] $BLIN[a, b] = B^{\sharp}[a, b]$

Definition 25.1 *If between sets $L \subset B[a,b]$ and $BK \subset BLIN[a,b]$ can be established one-to-one correspondence $K : BK \to L$ then*

1) BK is called the complete set,

2) K is called the complete mapping,

3) the sets BK and L are called mutually complete sets.

The functional

$$F[u] = \int_a^b f(x)u(x)dx \tag{25.2}$$

is the canonical form of representation of a linear complete functional.

If $L \subseteq B[a,b]$, then to every function $f \in L$ associate the linear functional (25.2).

Let $M \subset B[a,b]$. For a function $u \in M$ and a constant $\Delta > 0$ denote by U an extension of the function $u : [a,b] \to R$ on the interval $[a-\Delta, b+\Delta]$, i.e. $U : [a-\Delta, b+\Delta] \to R$, and $U(x) = u(x)$ for $x \in [a,b]$. $U_\zeta = \{U(\zeta+x), x \in [a,b]\}$ is the section of the ζ-shift[32] of the function U on the interval $[a-\zeta, b-\zeta]$.

Definition 25.2 *A function $U : [a-\Delta, b+\Delta] \to R$ ($\Delta > 0$) is the natural extension of a function $u \in M \subset B[a,b]$ in a class M if*

1) $U(x) = u(x)$, $x \in [a,b]$,

2) $U_\zeta = \{U(\zeta+x), x \in [a,b]\} \in M, |\zeta| < \Delta\}$.

The invariant derivative of a function $f \in L$ is defined by means of the invariant derivative of the corresponding linear functional (25.2).

For the convenience we formulate here the definition of the invariant derivative of the functional[33] (25.2).

Definition 25.3 *A linear functional (25.2) has at a function $u \in M \subset B[a,b]$ the invariant derivative $\partial F[u]$*

[32] $|\zeta| < \Delta$

[33] general definition of the invariant derivative see in the Part I.

in a class M, *if for every natural extension* U *in the class* M *the corresponding function*

$$\psi_U(\zeta) = F[U(\zeta + \cdot)] = \int_a^b f(x)U(\zeta + x)dx \qquad (25.3)$$

has at the zero the derivative $\dot{\psi}_U(0)$, *invariant with respect to natural extensions* U *of the function* u, *i.e. the value of the derivative* $\dot{\psi}_U(0)$ *is the same for all natural extensions* U. *The value of the invariant derivative is equal to* $\dot{\psi}_U(0)$:

$$\partial F[u] = \frac{d}{d\zeta}(\psi_U(\zeta))_{(\zeta=0)} = \frac{d}{d\zeta}\left(\int_a^b f(x)U(\zeta + x)dx\right)_{(\zeta=0)} \qquad (25.4)$$

A linear functional $F[u] : M \to R$ can be presented as the sum $F[u] = cF[u] + \hat{F}[u]$ of two linear functionals:

1) a linear singular functional $cF[u]$,

2) a linear regular (integral) functional $\hat{F}[u]$.

Definition 25.4 *A function* $w \in M$ *is the invariant derivative of a function* $f \in M$ *if the corresponding linear functional* $W[u] = \int_a^b w(x)u(x)dx$ *is the regular component* $\hat{\partial F}$ *of the invariant derivative* ∂F *of the linear functional*

$$F[u] = \int_a^b f(x)u(x)dx, \qquad (25.5)$$

That is

$$\hat{\partial F}[u] = W[u] = \int_a^b w(x)u(x)dx$$

For the invariant derivative of the function f we use the notation f' (similar to the classic derivative).

25.2 Examples

Consider a function $f \in L \subset B[a,b]$ and the corresponding linear functional (25.5).

To calculate the invariant derivative of the functional (25.5) at a function $u \in M$ fix a natural continuation U of the function u in a class M and consider the corresponding function (25.3).

According to the definition 25.3 the invariant derivative of the functional (25.5) is the derivative at the zero of the function (25.3), if this derivative

$$\frac{d}{d\zeta}(\psi_U(\zeta))_{(\zeta=0)} = \frac{d}{d\zeta}\left(\int_a^b f(x)U(\zeta+x)dx\right)_{(\zeta=0)} \qquad (25.6)$$

exists and is invariant [34] with respect to different extensions U.

On can obtain various types of the invariant derivatives of a function f choosing different classes of test functions M.

Example 25.1 Take as the test set M a subset $M^1[a,b]$ of the space of continuous differentiable functions. In this case a function $u \in M$ and its natural extensions U in the class M are continuous differentiable functions in domains. Then one can differentiate under the integral in the formula (25.6)

$$\frac{d}{d\zeta}(\psi_U(\zeta))_{(\zeta=0)} = \left(\int_a^b f(x)U'(\zeta+x)dx\right) =$$

$$= \left(\int_a^b f(x)u'(x)dx\right) \qquad (25.7)$$

here u' denotes the ordinary[35] derivative of the function. The expression (25.7) does not depend on specific exten-

[34] has the same value

[35] usual finite dimensional

sion U of the function u and, therefore, is the invariant derivative of the functional (25.5).

The functional (25.7) does not contain a singular component so $\hat{\partial F}[u] = \partial F[u]$.

Thus the linear on $C^1[a,b]$ functional (25.7), corresponding to the function f, is the invariant derivative of the linear functional (25.5), generated by the same function f. It follows that the invariant derivative of the function f (with respect to the class of continuous differentiable functions) is the original function[36] f.

Example 25.2 Let M be a subset of the space of continuous functions. If a function u and its natural extension U are only continuous functions then the function $\psi_U(\cdot)$ [see (25.3)] be differentiable at the zero if the function f be continuous differentiable. In this case one can differentiate under the integral in (25.6) and the derivative (25.6) can be calculated in the following way

$$\partial F[u] = \frac{d}{\zeta}(\psi_U(\zeta))_(\zeta = 0) = \frac{d}{d\zeta}(\int_a^b f(x)U(\zeta+x)dx)_{(\zeta=0)} =$$

$$= \frac{d}{d\zeta}(\int_{a+\zeta}^{b+\zeta} f(x-\zeta)U(x)dx)_{(\zeta=0)} =$$

$$= F(b)u(b) - f(a)u(a) + \int_a^b f'(x)u(x)dx\,. \qquad (25.8)$$

The value of this derivative is the same for all natural continuations U of the function u and, therefore, it is the invariant derivative. The regular component of the invariant derivative (25.8) has the form

$$\hat{\partial F}[u] = \int_a^b f'(x)u(x)dx\,.$$

[36] i.e. is equal to itself

It follows that the invariant derivative of a continuous differentiable function f (with respect to the class of continuous functions) is the function f' (i.e. its ordinary derivative).

Remark 25.1 The invariant derivative of a function is a global (on the whole domain), but not a local (at a point) characteristic of the function.

25.3 Relationship between the invariant derivative and the Sobolev generalized derivative

In the previous section we established that under appropriate choices of test functions M the invariant derivative of a function $f \in L$ can coincide with the function itself (this can be considered as the invariant differentiability of the zero order) or can be equal to the ordinary (usual) derivative of the function (that can be considered as the invariant differentiability of the first order).

In this section we study the relation of the invariant derivative with the Sobolev generalized derivative.

Denote by $CD[a, b]$ the space of infinite differentiable functions with supports in $[a, b]$. Further we show that if the set of the test functions is $M = CD[a, b]$, then the invariant derivative of a function f be equal to the Sobolev generalized derivative.

There exist several equivalent definitions (see, for example [29]) of the Sobolev generalized derivative. We will use the following

Definition 25.5 *A locally integrable function $w : [a, b] \to R$ is the generalized derivative of a locally integrable function $f : [a, b] \to R$, if for all $u \in CD[a, b]$*

$$\int_a^b f(x)u'(x)dx = \int_a^b w'(x)u(x)dx. \tag{25.9}$$

The following proposition is valid

Theorem 25.1 *If a function $f : [a,b] \to R$ is invariantly differentiable with respect to the class of functions $M \subset CD[a,b]$, then the function has the Sobolev generalized derivative which is equal to the invariant derivative.*

Proof. Let $w(\cdot)$ be the invariant derivative of the function $f : [a,b] \to R$ with respect to the class of functions $CD[a,b]$. According to the definition 25.4 it means that the linear canonic functional

$$F[u] = \int_a^b f(x)u(x)dx \tag{25.10}$$

has the invariant derivative $\partial F[u]$ $CD[a,b]$ with the regular component

$$\hat{\partial F}[\phi] = \int_a^b w(x)u'(x)dx \tag{25.11}$$

where $w : [a,b] \to R$ is a function from M.

Let us calculate the invariant derivative of the functional (25.10) by the definition 25.4. Fix an element $u \in CD[a,b]$ and its natural extension U. By virtue of smoothness of the function U the corresponding function

$$\psi_U(\zeta) = \int_a^b f(x)U(x)dx \tag{25.12}$$

is differentiable at the zero and

$$\psi'_U(0) = \int_a^b f(x)U'(x)dx = \int_a^b f(x)u'(x)dx \tag{25.13}$$

The derivative (25.13) does not depend on specific choice of extension U and, therefore, is the invariant derivative of the functional (25.11), that is

$$\partial F[u] = \int_a^b f(x)u'(x)dx. \tag{25.14}$$

On the other hand, the invariant derivative of the functional (25.10) has the form (25.11). Then from (25.11) and (25.14) follows

$$\int_a^b w(x)u(x)dx = \int_a^b f(x)u'(x)dx, \forall u \in CD[a,b]. \tag{25.15}$$

The equality (25.15) is valid for all $u \in CD[a,b]$, hence in view of the definition 25.5, the function w is the generalized derivative of the function f. Theorem is proved. □

The inverse statement is also valid.

Theorem 25.2 *If a function $f(x)$ has the Sobolev generalized derivative then it has the invariant derivative in the class $M = CD[a,b]$ and the generalized derivative is equal to the invariant derivative.*

Proof. Let a function $w(x)$ be the generalized derivative of the function $f(x)$. With the function f associate the linear canonic functional

$$F[\phi] = \int_a^b f(x)\phi(x)dx \tag{25.16}$$

with the domain $\phi \in CD[a,b]$. To calculate the invariant derivative of the functional (25.16) at an element $u \in CD[a,b]$ fix a natural extension U and consider the corresponding function (25.12). The function (25.12) allows differentiation under the integral with respect to the variable ζ, Then (25.13) is valid. Therefore the invariant derivative of the functional (25.16) is equal to (25.14). In view of the definition 25.5 the following equality is valid

$$\int_a^b f(x)u'(x)dx = \int_a^b w(x)u(x)dx.$$

Then the invariant derivative (25.14) is equal to

$$\partial F[u] = \int_a^b f(x)u'(x)dx = \int_a^b w(x)u(x)dx.$$

and is the linear canonical functional on $CD[a,b]$. From this by virtue of the definition 25.3 follows that the function f has the invariant derivative $w(x)$ with respect to the class $CD[a,b]$. □

Theorems 25.1 and 25.2 can be unified in one proposition

Theorem 25.3 *A function $f : [a,b] \to R$ has the Sobolev generalized derivative if and only if the function f has the invariant derivative with respect to the class of functions $M = CD[a,b]$. At the same time both derivatives are equal.*

26 Relation of the Sobolev generalized derivative and the generalized derivative of the distribution theory

26.1 Affinitivity of the generalized derivative of the distribution theory and the Sobolev generalized derivative

Affinitivity means existence of the mathematical relations between mathematical objects and/or structures.

In this section we consider relation between the generalized derivative of a functional[37] and the generalized derivative of a function[38].

From the very beginning of my study of the functional analysis and the distribution theory I tried to find the answer on the following

Question B. Does there exist a mathematical relation between the *generalized derivative* of distributions and the Sobolev *generalized derivative* of functions. Or this is just a terminological coincidence?

[37] in the sense of the distrubution theory

[38] the Sobolev generalized derivative

Nobody could gave me more or less acceptable explanation concerning the question and more than thirty years I tried to find an answer by myself.

During elaboration of i-smooth analysis it turn out well to establish the mathematical links between both generalized derivatives. Below we present the scheme which shows the relations between the generalized derivatives.

The generalized derivative of the distribution theory

$$\Updownarrow (A)$$

The invariant derivative of a functional

$$\Updownarrow (B)$$

The invariant derivative of a function

$$\Updownarrow (C)$$

The Sobolev generalized derivative of a function

1) (A) shows the relation of the distribution generalized derivative with the invariant derivative of a linear functional.

2) (B) indicates the relation of the invariant derivative of a linear functional with the invariant derivative of the generating function.

3) (C) displays equivalence of the invariant and generalized derivatives of functions.

26.2 Multiplication of generalized functions at the Hamel basis

Material of this section is not directly related to the invariant derivatives, nevertheless the described ideas arose during development of i-smooth analysis and possibly can be interesting for the ready, so the section is added as an appendix to the book.

In the section 8 of the first part (Chapter 1) we considered a nonlinear approach to the problem of distribution multiplication. In the framework of this approach the product of two distributions[39] is a nonlinear but infinitely smooth in the generalized sense functional. In this section the problem of distribution multiplication is considered in the framework of more general problem of defining a product which would be complete [40] in the space of linear functionals [41].

At that we proceed from the concept that multiplication operation is increasingly associated with the algebraic structure of the considered spaces, than with the topological properties.

For this reason further the principle role play some algebraic structures.

Multiplication of linear functionals

Let F be a locally convex space with the topology τ; $F^\sharp$ F' are algebraically and topologically adjoint to$F$ spaces respectively; $Kerf = \{\phi \in F | f(\phi) = 0\}$ is the Kernel of a functional $f \in F^\sharp$; $m[B]$ is the cardinal number of a set $B \subset F$.

Main algebraic structure that will be used for defining the multiplication operation is the Hamel basis of the space F. In F exists many Hamel basises. Fix one of them (generally speaking, arbitrary) and further will denote it by $H = \{h_\alpha\}_{\alpha \in A}$ (A is the index set).

Important property of the Hamel basis consists in the following: any element $\phi \in F$ can uniquely be presented in the form of a finite linear combinations of elements of the Hamel basis H:

$$\phi = \sum_{i=1}^{m} c_i h_{\alpha_i} \qquad (c_i \in R, i = \overline{1, m}) \tag{26.1}$$

For arbitrary f and $g \in F^\sharp$ define the product (denoting

[39] linear continuous functionals]

[40] closed

[41] closeness means that the product of two linear functionals be a linear functional.

by $\bigotimes$) according to the rule:

$$(f \bigotimes g)\phi = \sum_{i=1}^{m} c_i f(h_{\alpha_i}) g(h_{\alpha_i}) \tag{26.2}$$

Figuratibely the introduced multiplication operation consists of multiplication f and g at the Hamel basis H with the further redefinition by linearity[42] on the whole F.

It is easy to derify that for any $f \quad g \in F^\sharp$ the product $f \bigotimes g$ also belongs to $F^\sharp$, i.e. the space $F^\sharp$ is closed with respect to the product $\bigotimes$. Not so obvious the question about closeness of the topological space F' with respect to the product $\bigotimes$ because an answer naturally should depend on the topology τ of the space F'.

Note several properties of the introduced product. For $f \in F^\sharp$ $S_f = \{h \in H | f(h) \neq 0\}$ and the number $rank_H f = m[S_f]$ is called the rank of the functional f (at the Hamel basis H).

Theorem Γ **1.** If $f \in F^\sharp$ and $rank_H f < \infty$, then $f \in F'$.

Consequence Γ **1.** If for f and $g \in F'$ $\min\{rank_H f, rank_H g\} < \infty$, then $f \bigotimes g \in F'$ and $S_{f \bigotimes g} = S_f \bigcap S_g$.

Above we consider that the Hamel H basis is fixed. However generally speaking it can be chosen according to the multiplied functionals.

Theorem Γ **2.** For any finite set of functionals $\{f_1, \ldots, f_k\} \in F'$ there exists a Hamel basis H such that $f_i \bigotimes f_j \in F'$, $i, j = \overline{1, k}$.

In general case the question about topological closeness of the product $\bigotimes$ can be formulate in the form of the following

Problem Γ 1.

1) describe a locally convex topology $\tau^\star$ on F such that the product $\bigotimes$ is continuous and closed operation in F';

[42] with preserving the linearity property

2) is the topology of the point (a)invariant with respect to different Hamel basises.

Our main interest is to the problem of generalized functions multiplication. This case we will study in the next section

Multiplication in the space of generalized functions

In this section we consider multiplication at a Hamel basis in the space of tempered distributions S', $S = S(R)$ is the space of test functions on R.

Consider some properties of the product $f \bigotimes g$ $(f, g \in S')$ in case of $S_f = \{h\}$, i.e. $rank_H f = 1$.

In this case[43]: $f \bigotimes g \in S'$.

The following relations are valid:

$$(f \bigotimes g)' = (g, h) f'$$

and

$$F[f \bigotimes g] = (g, h) F[f]$$

Here the prime denotes the generalized derivative, and F is the Fourier transform. For n-th degree we have

$$(f^n)' = (f, h)^{n-1} f'$$

and

$$F[f^n] = (f, h)^{n-1} F[f]$$

Example 26.1 For the delta-function $\delta(x)$ take a Hamel basis H such that $S_{\delta(x)} = \{h\}$ (i.e. $rank\delta_H(x) = 1$). Then

$$\delta(x) \bigotimes g = (g, h)\delta(x) \sum_{i=1}^{m} c_i f(h_{\alpha_i}) g(h_{\alpha_i}) \qquad (26.3)$$

[43] by virtue of the corollary Γ 1.

In the particular case of the product of $\delta(x)$ and Heavyside function $\vartheta(x)$ the result of our multiplication (26.3) coincides with the result of [3]:

$$\vartheta(x)\delta(x) = c\delta(x),$$

However a constant c, interpreted in [3] as an arbitrary constant, in our case is determined explicitely: $c = (\vartheta(x), h)$ and, obviously depend on the choice of h.

Relation with other theories of multiplication generalized functions

One of the variants constructing a distribution multiplication theory can be based on the following approach: for every two multiplied distributions ($f, g \in S'$) select an appropriate[44] a Hamel basis. Abiding by this way further under $f \bigotimes g$ we understand multiplication at an a Hamel basis$H_{f,g}$, depending on the multiplied pair of distributions $\{f, g\}$. In the framework of such approach one can describe various variants of distribution multiplication.

Theorem 26.1 *Let L be a subset of S and in L a multiplication $*$ is defined. Then for arbitrary $f, g \in L$ (excepting the case $f * g = 0$ for $g = \lambda f \neq 0$) there exists a Hamel basis $H_{f,g}$ such that $f * g = f \bigotimes g$.*

Remark 26.1 From theorem 26.1 it follows that in the framework of the approach, based on Hamel basises, it is possible describe distributions multiplication theories in which $f * f \neq 0$ for $f \neq 0$.

[44]according to some parameters

Bibliography

[1] *Antosik P., Mikusinskii J., Sikorski R.* Theory of distributions. The sequential qpproach.Elsevier scientific publishing company. Amsterdam. 1973.

[2] *Bahvalov N.S.* Numerical Methods, Nauka, Moscow. 1973. (In Russian)

[3] *Bogolyubov N.N., Logunov A.A., Todorov I.T.* Fundamentals of the axiomatic approach to the quantum field theory. Moscow.: Nauka. 1969. (In Russian)

[4] *Colombeau J.F.* New Generalized Functions and Multiplication of Distributions. North-Holland, Amsterdam. 1984.

[5] *Colombeau J.F.* A Multiplication of Distributions // J. Math. Anal. Appl. 1983. Vol. 94. No. 1. pp. 96-115.

[6] *Cryer C., Tavernini L.* The numerical solution of Volterra functional differentional equations bu Euler's method // SIAM J. Numer. Anal. 1972. V. 9. pp. 105-129.

[7] *Dahlquist G.* Numerical integration of ordinary dfferential equations, // Math. Scand., 1956. Vol. 4, pp. 33-50.

[8] *Gelfand I.M., Shilov G.E.* The generalized functions. Moscow: Fizmatgiz. 1958. issue 1.

[9] *Hale J.K., Lunel S.M.V.* Introduction to fuctional differential equations.Springer-Verlag, New York; Heidelberg; Berlin. 1993. 280 p.

[10] *Hairer E., Norsett S.P., Wanner G.* Solving Ordinary Differential Equations. I: Nonstiff Problems. Springer, Berlin, 1987.

[11] *Hairer E., Lubich Ch.* Asymptotic expansion of the global error of fixed-step methods // Numer. Math. 1984. V. 45, No. 3. pp. 345-360.

[12] *Hairer E., Wanner G.* Solving Ordinary Differential Equations. II: Stiff and Differential-Algebraic Problems. Springer, Berlin, 1996.

[13] *G. Hall and J. Watt.* Modern numerical methods for ordinary differential equations, Clarendon Press, Oxford, 1976.

[14] *Ivanov V.K.* Hyperdistributions and Schwartz multiplication// Doklady of USSR Academy of sciences. 1972. V. 204, No. 5. pp. 1045-1049.

[15] *Ivanov V.K., Perminov V.V.* Nonlinear operators in convolutions: ordinary differential equations. Sverdlovsk: Ural university. 1989. (in Russian)

[16] *Kantorovich L.V., Akilov G.P.* Functional Analysis. Moscow. Nauka. 1977. 744 p. (in Russian)

[17] *Kim A.V.* The direct Lyapunov method in stability theory of systems with delays. Ekaterinburg: Ural university, 1992. (in Russian)

[18] *Kim A.V.* i-Smooth analysis and functional differential equations. Ekaterinburg. Russian Academy of Sciences. 1996. 236 p.

[19] *Kim A.V.* Functional differential equations. Application of i–smooth calculus. Kluwer Academic Publishers. The Netherlands. 1999. 165 P.

[20] *Skeel R.D.* Analysis of Fixed-Stepsize Methods // SIAM J. Numer. Anal. 1976. V. 13. pp. 664–683.

[21] *Kim A.V., Kwon W.H., Pimenov V.G., Han S.H., Lozhnikov A.B., Onegova O.V., Kang J.W.* Time-Delay System Toolbox (for use with MATLAB). User's guide. 2000.

[22] *Kim A.V., Pimenov V.G.* *i*-smooth analysis and numerical methods for functional differential equations. Regular and chaotic dynamics. Moscow-Izhevsk. 2004. 140 p.

[23] *Kim A.V., Pimenov V.G.* Numerical Methods for Time-delay Systems on the Basis of i-Smooth Analysis, Proceedings of the 15th World Congress on Scientc Computation, Modelling and Applied Mathematics. Berlin, August 1997. V. 1: Computational Mathematics, pp. 193-196.

[24] *Kim A.V., Pimenov V.G.* Multistep numerical methods for functional dfferential equations, Mathematics and Computers in Simulation, Vol. 45, pp. 377-384.

[25] *Kim A.V.* On the second Lyapunov method for systems with delays // Differential Equations. 1985. V. 21. No. 3. pp. 385–391.

[26] *Kim A.V.* Generalized Solutions of nonlinear differential equations, *Izv. VUZov: Matematika*, No. 5, pp. 58–63.

[27] *Kim A.V.* On the Problem of Multiplication of Linear Functionals and Distributions, *Izv. VUZov: Matematika*, No. 6, pp. 69-71.

[28] *Kolmogorov A.N., Fomin S.V.* Elements of theory of functions and functional analysis. M. Nauka. 1976. 542 p. (in Russian)

[29] *Nikolskii S.M.* Approximation of several variables functions and inclusion theorems. Moscow.: Nauka, 1977.

[30] *Nordsieck A.* On numerical integration of ordinary differential equations. Math. Comp. 1975. V. 16. pp. 22–49.

[31] *Samarskii A.A. and Gulin A.V.* Numerical methods. Moscow: Nauka, 1989. (in Russian)

[32] *Schwartz L.* Theorie des Distributions. I-II. Hermann, Paris. 1950.

[33] *Shvets A. Yu.* Influence of a variable time lag on the stability of oscillations of a pendulum with a vibrating suspension point // Ukrainian Mathematical Journal. 1985. V. 37. No. 1, pp. 127-129.

[34] *S.L.Sobolev.* Methode nouvelle a resoudre le probleme de Cauchy pour les equations lineaires hyperboliques normales. Matem. sbornik, 1936, V. 43:1, p. 39-72.

[35] *Sobolev S.L.* Some applications of functional analysis to mathematical physics.L. LSU. 1950. (in Russian)

[36] *Tavernini L.* One-step methods for the numerical solution of Volterra functional differentional equations // SIAM J. Numer. Anal. 1971. V. 8. pp. 789-795.

[37] *Volterra V.* Theorie Mathematique de la Lutte poir la Vie. Gauthier–Villars, Paris. 1931.

[38] *Volterra V.* Theory of Functionals and of Integral and Integro-Differential Equations. Moscow. Nauka. 1982. 304 pp. (in Russian)

[39] *Zverkina T.S.* Numerical integration of dfferential equations with delays. Transactions of the Seminar on Theory of Differential Equations with Deviating Arguments. 1975. Vol. IX, pp. 82 - 86.

Index

Also of Interest

Check out these other titles from Scrivener Publishing

Reverse Osmosis: Design, Processes, and Applications for Engineers 2nd Edition, by Jane Kucera, ISBN 9781118639740. This is the most comprehensive and up-to-date coverage of the "green" process of reverse osmosis in industrial applications, completely updated in this new edition to cover all of the processes and equipment necessary to design, operate, and troubleshoot reverse osmosis systems. *DUE OUT IN MAY 2015.*

Pavement Asset Management, by Ralph Haas and W. Ronald Hudson, with Lynne Cowe Falls, ISBN 9781119038702. Written by the founders of the subject, this is the single must-have volume ever published on pavement asset management. *DUE OUT IN MAY 2015.*

Open Ended Problems: A Future Chemical Engineering Approach, by J. Patrick Abulencia and Louis Theodore, ISBN 9781118946046. Although the primary market is chemical engineers, the book covers all engineering areas so those from all disciplines will find this book useful. *DUE OUT IN MARCH 2015.*

Fracking, by Michael Holloway and Oliver Rudd, ISBN 9781118496329. This book explores the history, techniques, and materials used in the practice of induced hydraulic fracturing, one of today's hottest topics, for the production of natural gas, while examining the environmental and economic impact. *NOW AVAILABLE!*

Formation Testing: Pressure Transient and Formation Analysis, by Wilson C. Chin, Yanmin Zhou, Yongren Feng, Qiang Yu, and Lixin Zhao, ISBN 9781118831137. This is the only book available to the reservoir or petroleum engineer covering formation testing algorithms for wireline and LWD reservoir analysis that are developed for transient pressure, contamination modeling, permeability, and pore pressure prediction. *NOW AVAILABLE!*

Electromagnetic Well Logging, by Wilson C. Chin, ISBN 9781118831038. Mathematically rigorous, computationally fast, and easy to use, this new approach to electromagnetic well logging does not bear the limitations of existing methods and gives the reservoir engineer a new dimension to MWD/LWD interpretation and tool design. *NOW AVAILABLE!*

Desalination: Water From Water, by Jane Kucera, ISBN 9781118208526. This is the most comprehensive and up-to-date coverage of the "green" process of desalination in industrial and municipal applications, covering all of the processes and equipment necessary to design, operate, and troubleshoot desalination systems. *NOW AVAILABLE!*

Tidal Power: Harnessing Energy From Water Currents, by Victor Lyatkher, ISBN 978111720912. Offers a unique and highly technical approach to tidal power and how it can be harnessed efficiently and cost-effectively, with less impact on the environment than traditional power plants. *NOW AVAILABLE!*

Electrochemical Water Processing, by Ralph Zito, ISBN 9781118098714. Two of the most important issues facing society today and in the future will be the global water supply and energy production. This book addresses both of these important issues with the idea that non-usable water can be purified through the use of electrical energy, instead of chemical or mechanical methods. *NOW AVAILABLE!*

Biofuels Production, Edited by Vikash Babu, Ashish Thapliyal, and Girijesh Kumar Patel, ISBN 9781118634509. The most comprehensive and up-to-date treatment of all the possible aspects for biofuels production from biomass or waste material available. *NOW AVAILABLE!*

Biogas Production, Edited by Ackmez Mudhoo, ISBN 9781118062852. This volume covers the most cutting-edge pretreatment processes being used and studied today for the production of biogas during anaerobic digestion processes using different feedstocks, in the most efficient and economical methods possible. *NOW AVAILABLE!*

Bioremediation and Sustainability: Research and Applications, Edited by Romeela Mohee and Ackmez Mudhoo, ISBN 9781118062845. Bioremediation and Sustainability is an up-to-date and comprehensive treatment of research and applications for some of the most important low-cost, "green," emerging technologies in chemical and environmental engineering. *NOW AVAILABLE!*

Sustainable Energy Pricing, by Gary Zatzman, ISBN 9780470901632. In this controversial new volume, the author explores a new science of energy pricing and how it can be done in a way that is sustainable for the world's economy and environment. *NOW AVAILABLE!*

Green Chemistry and Environmental Remediation, Edited by Rashmi Sanghi and Vandana Singh, ISBN 9780470943083. Presents high quality research papers as well as in depth review articles on the new emerging green face of multidimensional environmental chemistry. *NOW AVAILABLE!*

Energy Storage: A New Approach, by Ralph Zito, ISBN 9780470625910. Exploring the potential of reversible concentrations cells, the author of this groundbreaking volume reveals new technologies to solve the global crisis of energy storage. *NOW AVAILABLE!*

Bioremediation of Petroleum and Petroleum Products, by James Speight and Karuna Arjoon, ISBN 9780470938492. With petroleum-related spills, explosions, and health issues in the headlines almost every day, the issue of remediation of petroleum and petroleum products is taking on increasing importance, for the survival of our environment, our planet, and our future. This book is the first of its kind to explore this difficult issue from an engineering and scientific point of view and offer solutions and reasonable courses of action. *NOW AVAILABLE!*